Transport and Energy Processes 2017

Core Programming Area at the 2017 AIChE Annual Meeting

Minneapolis, Minnesota, USA
29 October - 3 November 2017

ISBN: 978-1-5108-5796-4

Printed from e-media with permission by:

Curran Associates, Inc.
57 Morehouse Lane
Red Hook, NY 12571

proceedings.com

Some format issues inherent in the e-media version may also appear in this print version.

Printed by Curran Associates, Inc. (2018)

For permission requests, please contact AIChE
at the address below.

AIChE
120 Wall Street, FL 23
New York, NY 10005-4020

Phone: (800) 242-4363
Fax: (203) 775-5177

www.aiche.org

Additional copies of this publication are available from:

Curran Associates, Inc.
57 Morehouse Lane
Red Hook, NY 12571 USA
Phone: 845-758-0400
Fax: 845-758-2633
Email: curran@proceedings.com
Web: www.proceedings.com

TABLE OF CONTENTS

Some abstracts in this book refer to images that were not made available by the publisher

505103 Session Keynote: Ionic Liquid Electrolytes for Electrochemical Energy Storage

Abstract:

Ionic Liquids (ILs) are liquid salts that have emerged as â€˜designer solventsâ€™ over the last twenty years due to their highly tunable chemical structure. Remarkable physical properties such as high thermal and electrochemical stability, non-flammability, and negligible volatility render ILs ideal electrolytes in electrochemical systems such as batteries and supercapacitors. Rate-capability and long-term cyclability of rechargeable batteries such as lithium-ion and sodium-ion batteries depend on many factors including the stability of the solid-electrolyte interface and ion conduction efficiency. As stand-alone electrolytes ILs are poor ion conductors because of their high viscosity, preferential ion interactions and ion aggregates that induce high-energy barriers for conduction. An increase in temperature often offsets some of the challenges associated with ion conduction. We have showed that a sodium-ion battery operating at 60 ËšC with ionic liquid electrolyte, Na[FSI] / [PYR13][FSI] (1-propyl-methylpyrrolidinium bis(fluorosulfonyl)imide), achieves superior performance in terms of capacity and cyclability, compared to organic carbonate-electrolytes at a current density of 100 mA/cm^2. Similarly, long-term cyclability with Li[TFSI] / [PYR13][TFSI] ([TFSI] is bis(trifluoromethylsulfonyl)imide) at 25 ËšC is achievable in Li-Li symmetrical cells, however at lower current densities (e.g., 0.1 mA/cm^2). In order to improve rates at which battery can discharge, while maintaining cyclability, we designed binary IL-IL mixtures as electrolytes with improved conductivities. Since ILs are variable in chemical makeup and in terms of stability against a metal such as the lithium metal, we investigated the solid-electrolyte interface to determine a pre-treatment strategy for the use of lithium metal as an electrode with IL electrolytes. A perspective on IL electrolytes for energy storage devices will be delivered in this presentation. Our most recent findings on the application of binary IL-IL mixtures in lithium batteries and the characteristics of the lithium-IL interfaces determined by Electrochemical Impedance Spectroscopy and X-ray Photoelectron Spectroscopy will be presented.

486464 A Novel Lithium Salt Additive for Improving Electrochemical Performance of High-Voltage $LiNi_{0.5}Mn_{1.5}O_4$ Cathode

Abstract:

Recently, with the raising concern on air pollution and greenhouse emission caused by the fossil fuel usage, lithium ion batteries, high energy density storages, become more attractive for transportation purpose like electric vehicles and hybrid electric vehicles [1]. To achieve this goal, the high voltage cathodes are intensively studied in recent years. Among various high voltage electrodes, $LiNi_{0.5}Mn_{1.5}O_4$ is expected to be the most promising one owing to its relatively moderate price and theoretical specific capacity (147mAh g^{-1}). However, the practical applications are limited because of transition metal dissolution and the severe oxidation decomposition of the commercial electrolyte caused by the unstable interface between the cathode and electrolyte under high voltage condition (>4.5V vs. Li^+/Li). There are several methods to improve the interface stability, such as coating the electrode surface, searching for more stable lithium salts and solvents, or employing additives. Among these methods, applying additives is the most efficient and economical way [2, 3].

In this short term, lithium fluorozirconate is synthesized [4] and added into the traditional electrolyte (1M $LiPF_6$ into EC, DEC, DMC which are mixed at the ratio of 1:1:1 by volume). Preliminary cycle performance tests show that adding lithium fluorozirconate can improve the discharge capacity of the $LiNi_{0.5}Mn_{1.5}O_4/Li$ half cells and improve the rate properties. Scanning electron microscope (SEM), high resolution transmission electron microscopy (TEM) and X-ray photoelectron spectroscopy (XPS) are performed to characterize the fresh and cycled electrodes with/without lithium fluorozirconate. SEM indicates that the cycled $LiNi_{0.5}Mn_{1.5}O_4$ electrode with lithium fluorozirconate maintains the clean and sharp morphology. As a comparison, without lithium fluorozirconate, the sharp edge of $LiNi_{0.5}Mn_{1.5}O_4$ electrode almost disappears completely after cycling. Furthermore, the results of TEM and EDS suggest that the cycled $LiNi_{0.5}Mn_{1.5}O_4$ electrode with lithium fluorozirconate displays a thin and uniform SEI layer which mainly consists of element zirconium. From the XPS observation, it can be inferred that the involvement of lithium fluorozirconate facilitates the formation of SEI layer and effectively inhibits the decomposition of both lithium salts and carbonate solvents, which is consistent with the results of TEM. And XPS depth analysis demonstrates that the SEI layer formed with lithium fluorozirconate can significantly suppress the dissolution of Mn, resulting in a better cycling performance of $LiNi_{0.5}Mn_{1.5}O_4/Li$ half cells.

References:

[1] Z.N.Wang, Y.J.Cai, Z.H.Wang, S.M.Chen, X.M.Lu, S.J.Zhang, J Solid State Electrochem 2013, 17, 2839â€"2848.

[2] K. Xu, Chem Rev 2014, 114, 11503-11618.

[3] Q. Li, J. Chen, L. Fan, X. Kong, Y. Lu, Green Energy & Environment 2016, 1, 18-42.

[4] S. K. Shi, [P].CN103227326A. 2013-07-31.

498810 *In situ* Generation of Stable Interphases in Lithium Oxygen Battery

Abstract:

Recent safety issues related to overcharging or fast-charging lithium-ion battery in addition to its limitations in terms of capacity and energy density have compelled researchers to look beyond lithium ion batteries. Replacing the graphite anode in Li-ion battery with a thin metallic lithium can result in significant improvements in multiple aspects like cost, weight and size. Perhaps the greatest advantage of this anodic replacement lies in the possibility of using of a Li-free high-capacity cathode like oxygen that can enhance the gravimetric energy density of battery from ~0.3kWh/kg to ~12kWh/kg (comparable to gasoline). However, these metal batteries, particularly the lithium-oxygen cell, have several fundamental roadblocks relating to both anode and cathode. The major anode specific problems are dendrite induced short circuits caused by uneven electrodeposition and rapid capacity fade due to internal side reactions. On the other hand, the oxygen cathode has several issues like low round trip efficiency, low specific capacity and poor rechargeability due to the insulating nature of the discharge product, lithium peroxide. In this study, we design ionomer based additives for forming *in-situ* solid-electrolyte interphases (SEIs) comprising of bromide salts and fixed anions that simultaneously addresses the issues of anode and cathode in a lithium-oxygen battery. The formation of these interphases was confirmed using scanning electron microscopy after focused ion milling at cryo temperatures as well as by chemical bond mapping using X-ray Photoelectron Spectroscopy. The ionomer SEIs protect the lithium anode against parasitic reactions and also stabilize the electrodeposition forming more compact and smooth lithium nucleates compared to that in usual batteries. The working mechanism of the ionomer additive is related to the role of bromide salts in reducing the energy barrier of ion transport at anode surface, as well as the effect of tethered anions in preventing the formation of a space charge region by depletion of free salts. Further, the bromine species, liberated in the ionomer anchoring reaction, act as a redox mediator in the oxygen evolution reaction (OER) at the cathode, thus reducing the charge overpotential. An added advantage of designing ionomer based SEI is the remarkable chemical stability it imparts to the lithium metal, enabling the usage of reactive electrolytes (like dimethyl sulfoxide and N,N dimethylacetamide), which also have very high-Gutmann donor number. These high donor number electrolytes have been reported to exhibit rare stability against nucleophilic attack by superoxides and at the same time they solvate intermediate cathode reaction products to impart high discharge capacity. We believe, these ionomer-based SEIs can rationally regulate the ion and mass transport in the lithium oxygen electrochemical cell to address the major barriers in lithium oxygen technologies.

499810 A Freestanding Calcium Hydroxide Interlayer As a Selective Separator for Rechargeable Alkaline Zn/MnO_2 Batteries

Abstract:

The alkaline Zn/MnO_2 battery has dominated the primary battery market since its invention because of its low cost and safety characteristics, as well as its high theoretical energy density that arises from the high gravimetric capacity of MnO_2. Currently, a long-term rechargeable Zn/MnO_2 battery is achievable at a reduced depth of discharge (DOD) (~10%-20% of the first electron of MnO_2) with a curtailment of energy density [1]. Its energy density higher than that of lead-acid batteries and the cost lower than $100/kWh make it a disruptive technology. Accessing higher DOD results in higher energy densities, but leads to detrimental characteristics like phase transformation of MnO_2 and the zinc redistribution [2][3]. Recently, a class of Bi-birnessite (a layered MnO_2 structure with bismuth oxide) material intercalated with Cu^{2+} has been reported as being able to stabilize the MnO_2 structure and deliver near-full two-electron capacity for more than 6,000 cycles in the absence of zinc [4]. However, when a zinc anode is employed, the most detrimental effect arises from zincate ions ($[Zn(OH)_4]^{2-}$) formed during discharge that traverse to the cathode side and react to form a spinel phase $ZnMn_2O_4$, a highly resistive and electrochemically inactive compound, leading to rapid energy density loss and battery failure [5][6]. Therefore, inhibiting $ZnMn_2O_4$ formation remains a major challenge for the long-term cycling of an energy dense secondary Zn/MnO_2 battery.

In this presentation, we report the use of an interlayer fabricated of calcium hydroxide for zincate sequestration, which successfully prevents $ZnMn_2O_4$ formation. $Ca(OH)_2$ is known to reversibly react with zincate ions and form an insoluble complex, calcium zincate ($Ca(OH)_2{\bullet}2Zn(OH)_2{\bullet}2H_2O$). Its role as an anode additive in Ni/Zn batteries is well known; however, its use has been limited due to its high resistivity and low density. We find that $Ca(OH)_2$ as an interlayer successfully avoids these drawbacks. It shows good properties such as a porous structure, good electrolyte retention ability, and a high ionic conductivity (Fig. 1a). It stabilizes the zincate concentration at a lower number in the bulk electrolyte by forming calcium zincate with excess zincate ions during discharge, and working as a zincate source during charge through decomposition (Fig. 1b). It is effective in localizing and trapping the zincate ions, while not affecting the transport of hydroxyl ions. More importantly, it inhibits the formation of zinc manganese spinel phase at high DOD as observed through X-ray diffraction (Fig. 2a) and energy dispersive X-ray spectroscopy analyses, which results in improved capacity retention and energy density in a Zn/MnO_2 cell (Fig. 2b).

References

[1] N. D. Ingale, J. W. Gallaway, M. Nyce, A. Couzis, and S. Banerjee, J. Power Sources, 276 (2015) 7-18

[2] D. E. Turney, J. W. Gallaway, G. G. Yadav, R. Ramirez, M. D'Ambrose, S. Kolhekar, M. Nyce, X. Wei, J. Huang, Y. Chen-Wiegart, J. Wang, S. Banerjee, Chemistry of Materials, under review.

[3] X. Wei, D. Desai, G.G. Yadav, D. E. Turney, A. Couzis, S. Banerjee, Electrochim. Acta, 212 (2016) 603-613

[4] G. G. Yadav, J. W. Gallaway, D. E. Turney, M. Nyce, J. Huang, X. Wei, S. Banerjee, Nat. Commun. 8, 14424 (2017)

[5] J. W. Gallaway, M. Menard, B. Hertzberg, Z. Zhong, M. Croft, L.A. Sviridov, D. E. Turney, S. Banerjee, D. A. Steingart, and C. K. Erdonmezg, J. Electrochem Soc., 162 (2015) A162-A168

[6] J. W. Gallaway, B. J. Hertzberg, Z. Zhong, M. Croft, D. E. Turney, G. G. Yadav, D. A. Steingart, C. K. Erdonmez, and S. Banerjee, J. Power Sources, 321 (2016) 135-142

Fig. 1 (a) Nyquist plots showing ionic conductivities of different separators measured by a symmetric cell; (b) Zincate ion concentrations in the cathode and anode chambers of a cycling cell at the end of each discharge half-cycle, with different separators applied.

Fig. 2 (a) XRD patterns of different EMD electrodes i) uncycled, ii) cycled vs. a Hg/HgO reference electrode, iii) cycled in a cell with a $Ca(OH)_2$ interlayer and iv) cycled in a $Ca(OH)_2$-free cell; (b) Curves of specific discharge capacity with cycle number.

501608 Regenerable Cu-Intercalated MnO_2 Layered Cathode for Highly Cyclable Energy Dense Aqueous-Based Batteries

Abstract:

Regenerable Cu-intercalated MnO_2 layered cathode for highly cyclable energy dense aqueous-based batteries

Batteries are important storage devices that can enable renewable energy integration into the grid. However, they need to meet certain requirements like safety, low cost, high cycle life and high energy densities for use in grid applications. Manganese dioxide (MnO_2)/Zinc (Zn) alkaline batteries have all the aforementioned desired attributes including very high volumetric energy densities of >400Wh/L (1) because of the high gravimetric capacities of its electroactive materials – 617mAh/g for MnO_2 and 820mAh/g for Zn based on 2 electron reactions. However, their use has been restricted to primary applications like common household appliances because rechargeability has been an issue due to the fundamental nature of MnO_2 and Zn (2). The cathode, MnO_2, has been known to undergo lattice expansion and crystal structure breakdown beyond a certain depth-of-discharge (DOD). Prior work has concentrated on limiting the DOD of MnO_2 (5-10% of 617mAh/g) to make the battery rechargeable (3); however, this has resulted in low energy densities. Accessing the 2nd electron capacity (617mAh/g) of MnO_2 reversibly has been the holy grail of MnO_2/Zn batteries to obtain very high energy dense batteries for use as low cost safe aqueous batteries in grid and electric vehicle applications (4).

In this presentation, we report the breakthrough of reversibly accessing the 2nd electron capacity of MnO_2 by using its layered polymorph called birnessite mixed with bismuth oxide (Bi-birnessite) and intercalating the layers with Cu ions (5). Bi-birnessite undergoes conversion reactions in alkaline electrolyte and ultimately forms electro-inactive hausmannite (Mn_3O_4) because of its poor charge transfer characteristics. Intercalating the layers of Bi-birnessite with Cu ions is shown to improve its charge transfer characteristics dramatically and regenerate its layered structure reversibly for thousands of cycles. We also present a case of Cu-intercalated Bi-birnessite's applicability in practical batteries by cycling the material at high areal capacities (10-29mAh/cm^2) for thousands of cycles at C-rates that are of interest in the battery community. Finally, a Cu-intercalated Bi-birnessite/Zn battery is shown to reversibly cycle at 140 Wh/L for over 90 cycles.

Figure 1. (a) Volumetric energy density of a Cu-intercalated Bi-birnessite/Zn cell. Inset shows the first 5 cycles of the cell. (b) Energy density comparison of different energy storage systems. (c) Specific capacity (mAh/g) vs cycle number for the Cu-intercalated Bi-birnessite against a sintered Ni counter electrode. Insets show specific cycle discharge curves for different wt.% loadings of MnO_2.

References:

1] Gallaway, J. W.; Hertzberg, B. J.; Zhong, Z.; Croft, M.; Turney, D. E.; Yadav, G. G.; Steingart, D. A; Erdonmez; C. K.; Banerjee, S. *"Operando identification of the point of [Mn_2]O_4 spinel formation during γ-MnO2 discharge within batteries"* Journal of Power Sources **321,** 135-142 (2016).

2] Wei, X.; Desai, D.; Yadav, G. G.; Turney, D. E.; Couzis, A.; Banerjee, S. *"Impact of anode substrates on electrodeposited zinc over cycling in zinc-anode rechargeable alkaline batteries"*. Electrochimica Acta **212**, 603-613 (2016)

3] Ingale, N. D.; Gallaway, J. W.; Nyce, M.; Couzis, A.; Banerjee, S., *"Rechargeability and economic aspects of alkaline zincmanganese dioxide cells for electrical storage and load leveling,"* Journal of Power Sources **276**, 718 (2015)

4] Dzieciuch, M. A.; Gupta, N.; Wroblowa, H. S. "*Rechargeable cells with modified MnO_2 cathodes.*" J. Electrochem. Soc. **135**, 2415–2418 (1988)

5] Yadav, G. G.; Gallaway, J. W.; Turney, D. E.; Nyce, M.; Huang, J.; Wei, X.; Banerjee, S. *"Regenerable Cu-intercalated MnO_2 layered cathode for highly cyclable energy dense batteries"* Nat. Commun. **8**, 14424 (2017).

500316 Investigation of Redox Active Oligomers for Nonaqueous Flow Batteries

Abstract:

Redox flow batteries are promising electrochemical technologies for grid-scale energy storage due to their many favorable inherent attributes including a decoupled power and energy rating, long operational lifetimes, and simple manufacturing[1,2]. However, current start-of-the-art flow batteries, which utilize acidic aqueous electrolytes and transition metal compounds, are too expensive to meet the U.S. Department of Energy's long range cost targets[3], spurring research into next-generation chemistries that can reduce battery system costs through lower materials costs and improved performance. Recently, RFBs leveraging nonaqueous electrolytes and organic active species have gained traction[2]. Transitioning to nonaqueous electrolytes enables wider electrochemical windows and access to a broader palette of candidate active materials. Redox active organics materials are of particular interest as their molecular structure can be tuned to impart desired electrochemical and physical properties.[4] Moreover, many candidate active materials may be produced at low costs from abundant feedstocks. Thus, taken together, this approach enables lower system costs through increased energy density.

Recent work has suggested that the use of redox active macroarchitectures, such as redox active oligomers (RAOs)[5], polymers (RAPs)[6], and colloids (RACs)[7], could significantly reduce RFB costs by enabling inexpensive size-selective separators to be used in place of fluorinated ion-exchange membranes. However, shifting from small molecules (monomers) to macrostructures has potentially significant, yet poorly understood effects on electrode kinetics, species transport, and electrolyte resistance; all of which can impact the technical performance and economic viability of the battery system. To begin to deconvolute these effects, we examine the electrochemical and solution properties of a series of electrolytes based on different sized RAOs using a suite of electrochemical, flow cell, and analytical techniques. Insights gained from these studies inform design decisions for next-generation active species and operating envelopes for RFBs based on these materials.

(1) Weber, A. Z.; Mench, M. M.; Meyers, J. P.; Ross, P. N.; Gostick, J. T.; Liu, Q. Redox Flow Batteries: A Review. *J. Appl. Electrochem.* **2011**, *41* (10), 1137–1164.

(2) Su, L.; Kowalski, J. A.; Carroll, K. J.; Brushett, F. R. Recent Developments and Trends in Redox Flow Batteries. In *Rechargeable Batteries*; Zhang, Z., Zhang, S. S., Eds.; Green Energy and Technology; Springer International Publishing, 2015; pp 673–712.

(3) Akhil, A. A.; Huff, G.; Currier, A. B.; Kaun, B. C.; Rastler, D. M.; Chen, S. B.; Cotter, A. L.; Bradshaw, D. T.; Gauntlett, W. D. DOE/EPRI 2013 Electricity Storage Handbook in Collaboration with NRECA. *Ed Albuq. NM Sandia Natl. Lab.* **2013**.

(4) Kowalski, J. A.; Su, L.; Milshtein, J. D.; Brushett, F. R. Recent Advances in Molecular Engineering of Redox Active Organic Molecules for Nonaqueous Flow Batteries. *Curr. Opin. Chem. Eng.* **2016**, *13*, 45–52.

(5) Doris, S. E.; Ward, A. L.; Baskin, A.; Frischmann, P. D.; Gavvalapalli, N.; Chénard, E.; Sevov, C. S.; Prendergast, D.; Moore, J. S.; Helms, B. A. Macromolecular Design Strategies for Preventing Active-Material Crossover in Non-Aqueous All-Organic Redox-Flow Batteries. *Angew. Chem. Int. Ed.* **2017**, *56* (6), 1595–1599.

(6) Nagarjuna, G.; Hui, J.; Cheng, K. J.; Lichtenstein, T.; Shen, M.; Moore, J. S.; Rodríguez-López, J. Impact of Redox-Active Polymer Molecular Weight on the Electrochemical Properties and Transport across Porous Separators in Nonaqueous Solvents. *J. Am. Chem. Soc.* **2014**, *136* (46), 16309–16316.

(7) Montoto, E. C.; Nagarjuna, G.; Hui, J.; Burgess, M.; Sekerak, N. M.; Hernández-Burgos, K.; Wei, T.-S.; Kneer, M.; Grolman, J.; Cheng, K. J.; Lewis, J. A.; Moore, J. S.; Rodríguez-López, J. Redox Active Colloids as Discrete Energy Storage Carriers. *J. Am. Chem. Soc.* **2016**, *138* (40), 13230–13237.

499305 An Analytical Model to Optimize the Cost-Effectiveness of Flow Batteries for the Electrical Grid

Abstract:

The world continues to modernize its electrical grid with more renewable sources, among which wind and solar photovoltaics are fastest growing[1]. While these sectors are becoming more competitively priced[2], their grid penetration is still limited by the variability of their supply. Secondary energy storage would then go far to enable renewable power, although such energy storage must likewise be cost-effective.

Redox flow batteries (RFBs) reversibly store energy in solubilized redox couples, which are stored external to the electrodes. This allows independent design of power and energy capacity and avails RFBs to a range of applications. The design space is further widened by an influx of chemical, geometric and operational innovations. Other advantages of RFBs include fast response, good efficiency and long cycle life.

While RFBs are getting closer to market, their grid-scale costs are confounded by the abundance of optimizable design conditions; moreover, experimental scale-up studies can exhaust a great deal of time and capital. A unified model, on the other hand, could quickly predict performance and cost characteristics of candidate RFBs, provided it be generalizable and contain an optimization framework to distill design variables into the lowest cost basis. We detail such a model[3] and validate its robustness for several state-of-the-art RFBs **(Figure 1)**. Our model could direct future efforts toward the most technically and economically promising designs **(Figure 2)**, ushering in a new pace for RFB development.

Figure 1. Model validation with charge discharge data for the **(a)** all-vanadium[4], **(b)** polysulfide-bromide[5], **(c)** polysulfide-iron[6], and **(d)** zinc-iron[7] batteries.

Figure 2. Modeled **(a)** battery capital costs, **(b)** installed capital costs, **(c)** cell efficiencies, and **(d)** stack efficiencies for zinc-ferricyanide and various state-of-the-art RFBs[3-7].

References:

1. International Energy Agency. *World Energy Outlook 2016*. **2016**; WEO-2016.
2. Yang Z, Zhang J, Kintner-Meyer MCW, Lu X, Choi D, Lemmon JP, Liu *J. Chem. Rev.* **2011**; 111:3577.
3. Ma RX *et al*. In preparation.
4. You D, Zhang H, Chen J. *Electrochim. Acta* **2009**; 54:6827.
5. Zhao P, Zhang HM, Zhou HT, Yi BL. *Electrochimica Acta* **2005**; 51:1091.
6. Gu S, Gong K, Yan EZ, Yan YS. *Energy Environ. Sci.* **2014**; 7:2986.
7. Gong K, Ma RX, Conforti KM, Kuttler KJ, Grunewald JB, Yeager KL, Bazant MZ, Gu S, Yan YS. *Energy Environ. Sci.* **2014**; 7:2986.

Dynamic Modeling of High-Temperature Sodium Sulfur batteries

Sai Pushpitha Vudata, Debangsu Bhattacharyya, Richard Turton
Chemical and Biomedical Engineering Department
West Virginia University

Overview

- Introduction
- Motivation
- Model Development
- Dynamic Studies
- Reduced Order Modeling
- Battery Development
- Air Cooling
- Results and Discussions
- Conclusion
- References

Introduction

- NaS battery proposed by Ford Motor Company (1968)
  - Molten sodium and sulfur electrodes
  - Beta"-alumina electrolyte (β''-Al_2O_3)
- Half-cell, overall reactions:

$$2Na \leftrightarrow 2Na^{+} + 2e^{-}$$

$$2Na^{+} + xS + 2e^{-} \leftrightarrow Na_2S_x$$

$$2Na + xS \leftrightarrow Na_2S_x$$

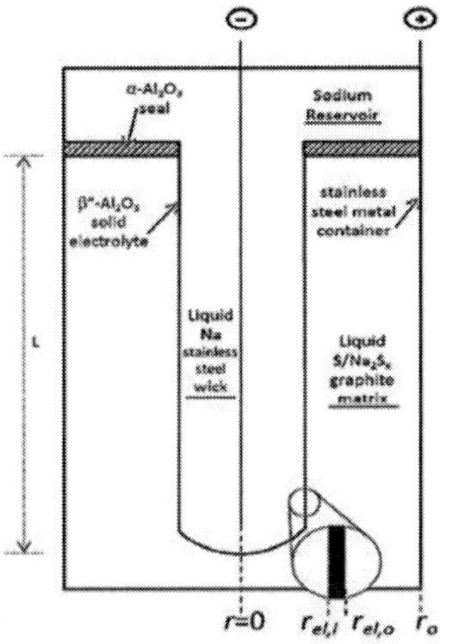

Schematic of a Sodium Sulfur Cell

WestVirginiaUniversity.

"Performance and Thermal Behavior of Sodium-Sulfur Cell under High Current Density Operations". Hiroyuki Kawamoto. *J. Electrochem. Soc.*, Vol. 136, No. 5, May 1989

Motivation

- To smoothen out fluctuations at grid level for load following applications
  - Large number of cells
  - High current density
  - Rapid cycling operation
  - Thermal management
- NaS battery is an ideal candidate for large-scale power storage
  - High energy density
  - Low cost of reactants
  - High open circuit voltage
  - Compact – 1/3 area for installation of Lead Acid

WestVirginiaUniversity.

Model Development

Approach and Objectives

- Sodium electrode:
  - Stainless steel wick
  - Complete wetting of electrolyte surface
  - Half-cell reaction occurs at surface of electrolyte only
- Beta"-alumina electrolyte:
  - Transporter of Na^+ ions only
- Sulfur electrode:
  - Graphite matrix
  - Half-cell reaction

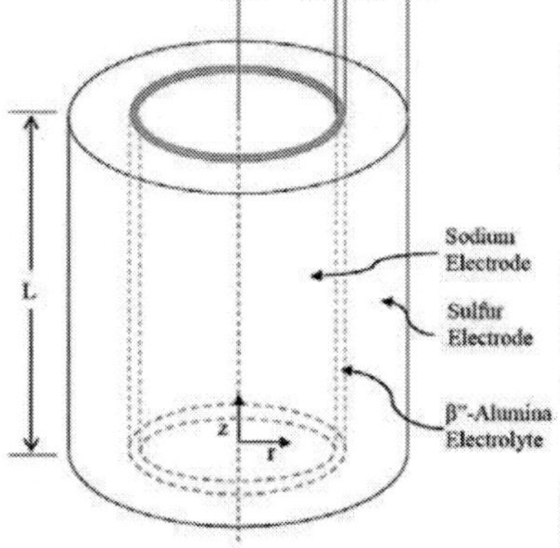

Simple representation of a sodium sulfur cell

WestVirginiaUniversity.

"Performance and Thermal Behavior of Sodium-Sulfur Cell under High Current Density Operations". Hiroyuki Kawamoto. *J. Electrochem. Soc.*, Vol. 136, No. 5, May 1989

Governing Equations: Sulfur Electrode

- Conservation of sodium ions

$$\frac{\partial C_{+(s)}}{\partial t} = -\frac{1}{F}\frac{a_s}{K_p}\left(\varphi_{e(s)} - \varphi_{i(s)} - \mu\right)$$

- Potential relations

$$-\sigma_{i(s)}\frac{\partial}{\partial r}\left(r\frac{d\varphi_{i(s)}}{dr}\right) + r\frac{a_s}{K_p}\varphi_{i(s)} = \frac{1}{3}FD_{(s)}\left(\frac{3t_+ - 1}{t_+(1-t_+)}\right)\frac{\partial}{\partial r}\left(r\frac{dC_{+(s)}}{dr}\right) + r\frac{a_s}{K_p}\left(\varphi_{e(s)} - \mu\right)$$

$$-\sigma_{e(s)}\frac{\partial}{\partial r}\left(r\frac{d\varphi_{e(s)}}{dr}\right) + r\frac{a_s}{K_p}\varphi_{e(s)} = r\frac{a_s}{K_p}\left(\varphi_{i(s)} + \mu\right)$$

WestVirginiaUniversity.

Governing Equations: Sulfur Electrode

- Energy balance

$$r\rho_{(s)}C_{p(s)}\frac{\partial T_{(s)}}{\partial t}-\frac{\partial}{\partial r}\left(r\lambda_{(s)}\frac{\partial T_{(s)}}{\partial r}\right)=rq_{(s)}$$

- Heat source term

$$q_{(s)}=Q_{Ohmic}+Q_{Entropy}$$

$$Q_{Ohmic}=\frac{1}{\sigma_{e(s)}}\left(\frac{I}{2\pi rL}\right)^2$$

$$Q_{Entropy}=\frac{a_s}{K_p}\left(\varphi_{e(s)}-\varphi_{i(s)}-\mu\right)\underbrace{\left(T\frac{dE}{dT}\right)}_{\text{Entropy term [Knoedler, 1984]}}$$

- Local potential

$$\mu=\begin{cases}2.05, & M\geq 0.710286\\ \alpha_1+\beta_1X_e, & M\leq 0.710286\end{cases}$$

WestVirginiaUniversity.

Governing Equations: Beta"-Alumina

- Sodium ion conservation

$$\frac{\partial C_{+(el)}}{\partial t}=\frac{D_{(el)}}{p_{(el)}}\frac{1}{r}\frac{\partial}{\partial r}\left(r\frac{\partial C_{+(el)}}{\partial r}\right)$$

- Potential for positive ions

$$-\sigma_{i(el)}\frac{\partial\varphi_{i(el)}}{\partial r}-FD_{(el)}\frac{\partial C_{+(el)}}{\partial r}=J$$

- Energy balance

$$\frac{\partial T_{(el)}}{\partial t}=\frac{1}{\rho_{(el)}C_{p(el)}r}\left(q_{(el)}r+\frac{\partial}{\partial r}\left(r\lambda_{(el)}\frac{\partial T_{(el)}}{\partial r}\right)\right)$$

  - Heat source term

$$q_{(el)}=\frac{J^2}{\sigma_i}$$

WestVirginiaUniversity.

Governing Equations: Sodium Electrode

- Sodium ion concentration at electrolyte interface

$$\frac{\partial C_{+(Na)}}{\partial r} = \frac{-I}{FV_{(Na)}}$$

- Ionic potential drop from sodium electrode to electrolyte interface

$$\varphi_{i(Na)} = -\eta_{(Na)} = (IRT_{(Na)})/(2\pi r_{el,i} L i_0 F)$$

- Energy balance

$$m_{ss} C_{p,ss} \frac{\partial T_{(Na)}}{\partial t} + m_{Na} C_{p,Na} \frac{\partial T_{(Na)}}{\partial t} = q_{(Na)} + 2\pi r_{el,i} L \lambda_{(el)} \left.\frac{\partial T_{(el)}}{\partial r}\right|_{r=r_{el,i}}$$

  - Heat source term

$$q_{(Na)} = \frac{I^2 \rho_{L(Na)} r_{el,i}}{2\pi \left({r_{el,i}}/{2} \right) L}$$

West Virginia University

Dynamic Simulation

- System of PDEs
- Aspen Custom Modeler
  - Method of lines
- Produce: transient profiles
- Validated by experimental data*
  - calculated trends were generated with the model, and the measured data on these figures are experimental data*.

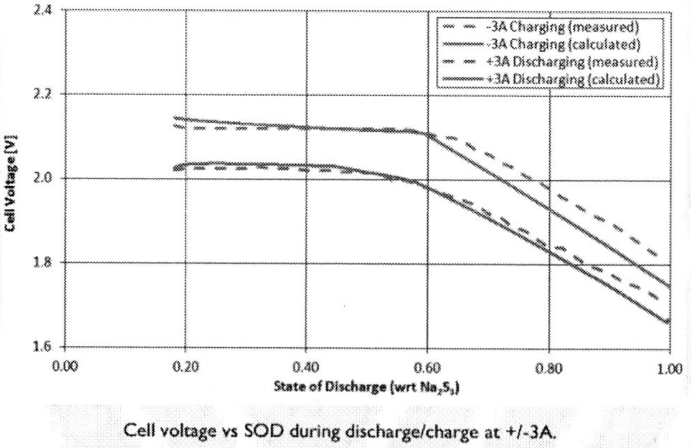

Cell voltage vs SOD during discharge/charge at +/-3A.

West Virginia University

"Performance and Thermal Behavior of Sodium-Sulfur Cell under High Current Density Operations". Hiroyuki Kawamoto. *J. Electrochem. Soc.*, Vol. 136, No. 5, May 1989

Temperature profile

- Spatial profile of temperature during charge and discharge with corresponding state of discharge

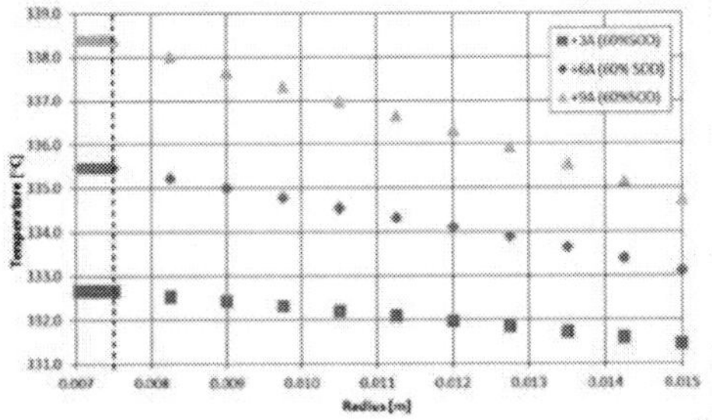

Temperature profile over cell during discharge at 60% SOD

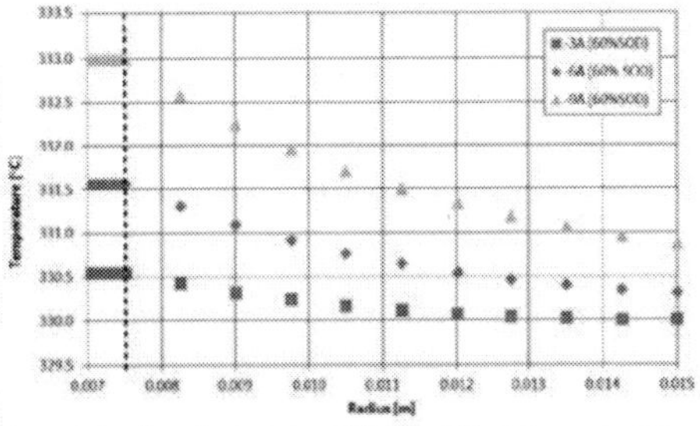

Temperature profile over cell during charge at 60% SOD

Sensitivity Studies

- During discharge, the ambient temperature has stronger effect on voltage profile with respect to state of discharge
- Cell voltage profile strongly depends on the current density and therefore can be used to regulate the voltage profile especially during charging

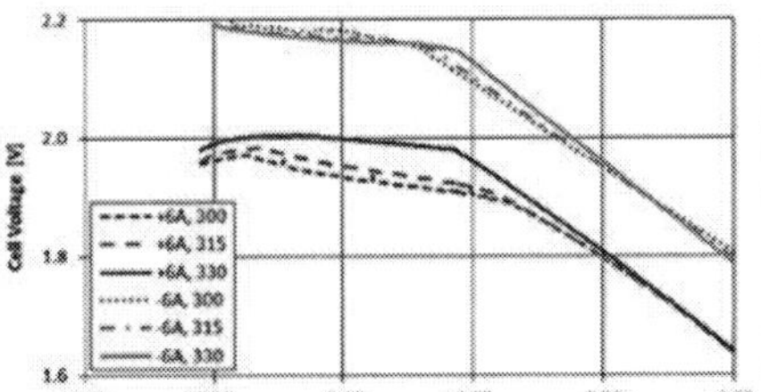

Cell voltage vs SOD during discharge and charge for varying ambient temperature

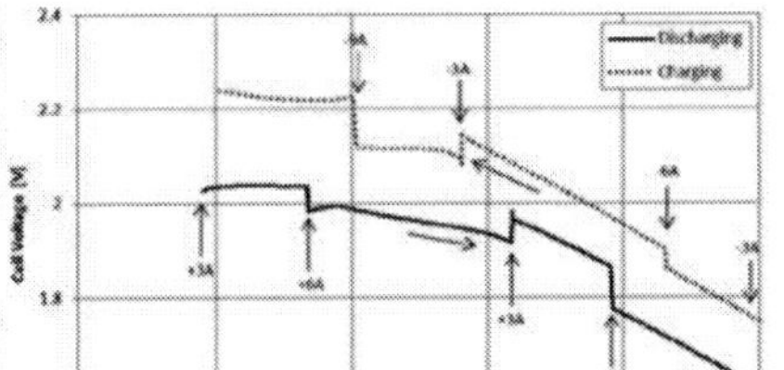

Cell voltage vs SOD during discharge and charge with current step changes

Reduced Order Modeling

- PDE-based model is
  - Computationally expensive for a battery comprising of thousands of cells.
  - It takes long time- Large number of DAEs to solve
  - Full order spatial discretization is slow
- Reduced order modeling using **Orthogonal collocation**
  - Better computational speeds
  - More accurate
  - Fewer node points

West Virginia University

Reduced Order Modeling

- Reduced order modeling using **Orthogonal collocation**:
  - **Coordinate Transformation**: Each region of the cell is rescaled to a domain of [0, 1].
  - Then the following equation is used:

$$u(R,t) = F(R,t) + \sum_{k=0}^{N} B_k(t)\,T_k(R)$$

  - The best possible approximation is found by choosing the collocation points as zeros of a specific class of Jacobi polynomials.

→

- ➢ System of ODEs
- ➢ 3 collocation points
- ➢ Aspen Custom Modeler V.8.4.
- ➢ Compared with the PDE model

West Virginia University

"Coordinate Transformation, Orthogonal Collocation, Model Reformulation and Simulation of Electrochemical-Thermal Behavior of Lithium-Ion Battery Stacks" Paul W. C. Northrop, Venkatasailanathan Ramadesigan, *Journal of The Electrochemical Society*

Battery Development

- Ambient temperature around single cell is considered to be varying
- Instances of a single cell model are developed
- Module- 20 cells in 10series/2parallel configuration
- All cells are simulated simultaneously
- The inlet temperature of a cell is assumed to be the exit temperature of preceding cell

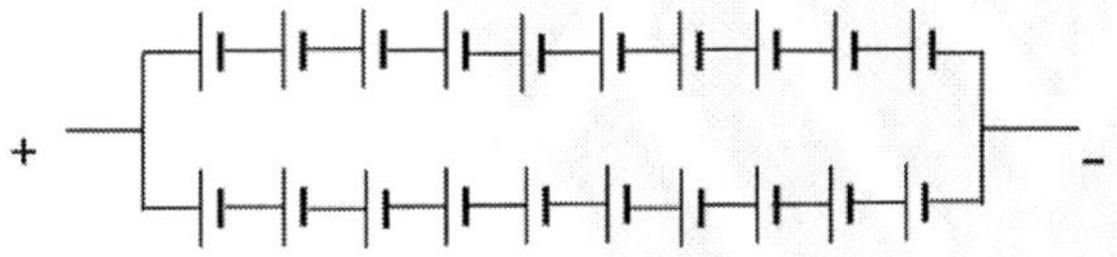

10series/2parallel configuration

Air Cooling

- Advantages
  - low viscosity
  - light weight of air
  - low cost
  - easy maintenance of the cooling system
- Energy conservation equation of air

$$\rho_a C_{p(air)} \frac{\partial T_{air}}{\partial t} + v \frac{\partial T_{air}}{\partial R} = \sum q_{cell}$$

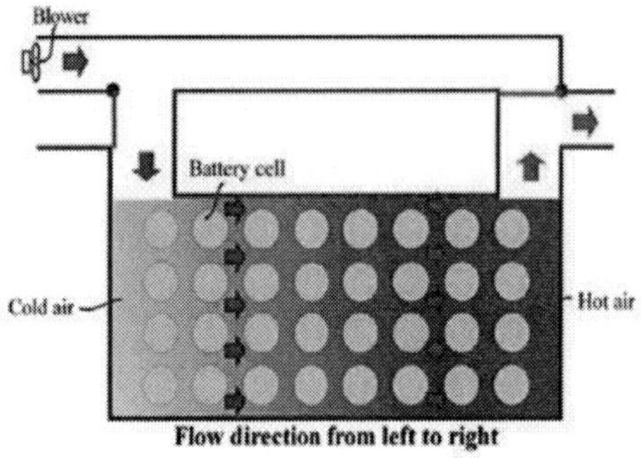

Flow direction from left to right

Results and Discussions

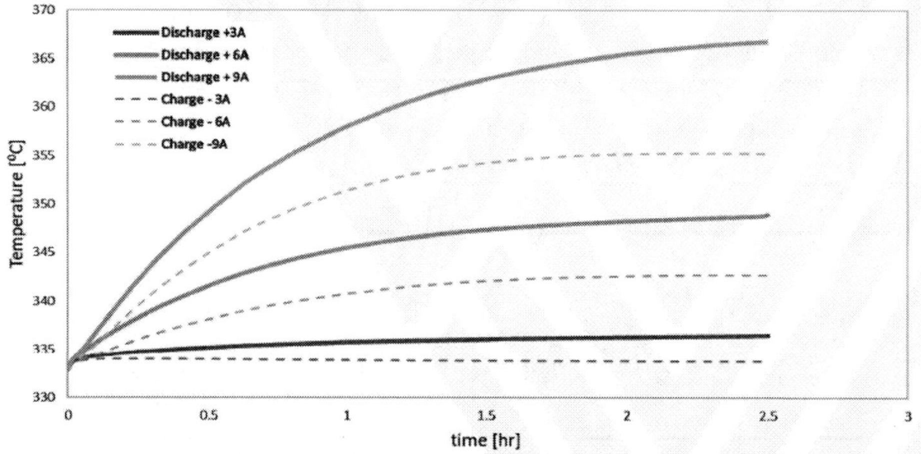

WestVirginiaUniversity.

17

Results and Discussions

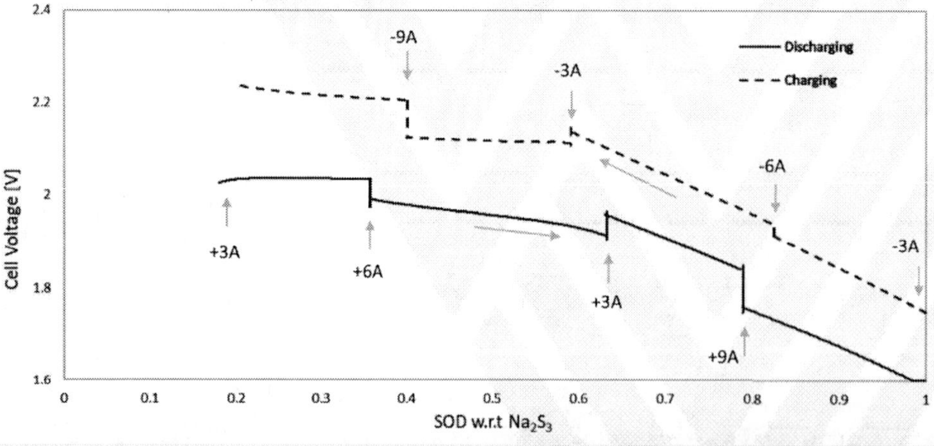

WestVirginiaUniversity.

18

Results and Discussions

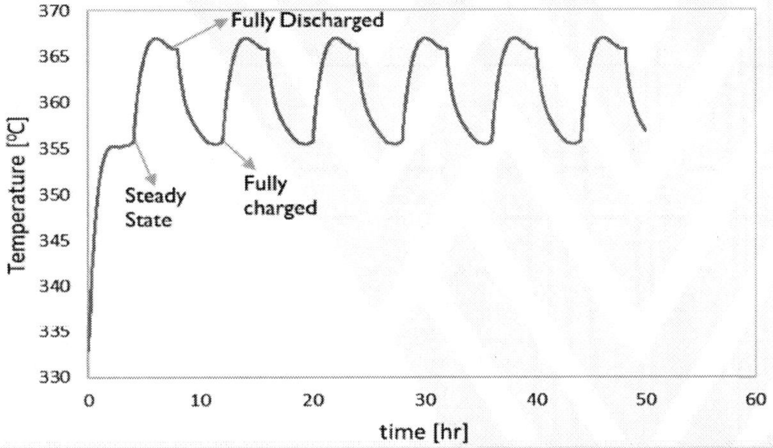

West Virginia University

Conclusions

- Sodium Sulfur cell (beta"-alumina electrolyte)
- Dynamic model
  - Non-isothermal
  - Transient
  - Each part of cell
- Results agree with experimental data
- Computational time decreased by converting PDE-based model to ODE-based model
- Thermal management system with air cooling

West Virginia University

References

- M. **McKubre**, S. Smedley and F. Tanzella, "The electrochemical Impedance of the Na/S Cell I. Experiments and Results," *Journal of the Electrochemical Society, vol. 136, no. 7, pp. 1962-1968, 1989.*
- H. **Kawamoto**, "Dynamic simulation of the charge-discharge characteristics of the sodium-sulphur cell," *Journal of Applied Electrochemistry, vol. 21, pp. 409-414, 1991.*
- R. **Knoedler**, "Calorimetric determination of the heat generation rate of sodium-sulfur cells during discharge and charge," *Journal of the Electrochemical Society, vol. 131, no. 4, pp. 845-850, 1984.*
- R. **Knoedler**, "Thermal properties of sodium-sulfur cells," *Journal of Applied Electrochemistry, vol. 14, pp. 39-46, 1984*
- H. **Kawamoto**, M. Wada and H. Hatoh, "Heat transfer characteristics of a sodium-sulfur cell," *Denki kagaku, vol. 54, no. 2, pp. 134-138, 1986.*
- H. **Kawamoto** and Y. **Kusakabe**, "Performance and thermal behavior of sodium-sulfur cell under high current density operations," *Journal of the Electrochemical Society, vol. 136, no. 5, pp. 1355-1361, 1989.*
- J. **Min** and C. **Lee**, "Numerical study on the thermal management system of a molten sodium-sulfur battery module," *Journal of Power Sources, vol. 210, pp. 101-109, 2012.*
- J. **Sudworth** and A. **Tilley**, The Sodium Sulfur Battery, New York: Chapman and Hall, 1985.
- "Coordinate Transformation, Orthogonal Collocation, Model Reformulation and Simulation of Electrochemical-Thermal Behavior of Lithium-Ion Battery Stacks" **Paul W. C. Northrop**, **Venkatasailanathan Ramadesigan,** *Journal of The Electrochemical Society*, **158** (12) A1461-A1477 (2011)

West Virginia University

21

Acknowledgment

U.S. DEPARTMENT OF ENERGY

U.S. - China Clean Energy Research Center
中美清洁能源研究中心

West Virginia University

22

496632 Cycle Life Study and Failure Analysis of the Rechargeable Porous Zinc Electrode in Alkaline Electrolyte

Abstract:

Metallic zinc (Zn) is an attractive negative electrode material for rechargeable batteries because it is a non-toxic, earth abundant metal. It has a low equilibrium potential of -1.35 V vs. mercury-mercuric oxide (Hg/HgO) in concentrated alkaline electrolyte and a high gravimetric charge capacity of 820 mAh/g. During discharge, Zn oxidizes and forms a dissolved complex known as the zincate ion, $Zn(OH)_4^{2-}$, from which zinc oxide (ZnO) precipitates as a solid. On charge, $Zn(OH)_4^{2-}$is reduced to metallic Zn. Motivation to study the Zn electrode in rechargeable batteries is driven by a target overall battery energy density of 200 Wh/L that is sustainable for 1000 or more charge-discharge cycles. This corresponds to a Zn electrode charge density in the range of 650 mAh/mL.

The percentage of charge capacity of Zn accessed is known as the depth of discharge (DOD). Battery energy density of 200 Wh/L motivates zinc electrode improvement to 20% Zn DOD. However, increased Zn DOD leads to a decrease in cycle life. A negative exponential relationship between cycle life and Zn DOD for the range of 1% to 15% Zn DOD for zinc-manganese dioxide ($Zn-MnO_2$) batteries was observed experimentally. The effect of additives on the Zn electrode at Zn DOD values in the range of 15% to 30% was studied in zinc-nickel (Zn-Ni) battery cycle tests. Additives used were surfactant cetyltrimethylammonium bromide (CTAB), bismuth oxide (Bi_2O_3), synthetic layered silicate Laponite, and calcium hydroxide ($Ca(OH)_2$). A cycle life of 542 at 15% Zn DOD was observed for a Zn electrode containing CTAB, bismuth oxide, and Laponite. In both $Zn-MnO_2$and Zn-Ni batteries, absolute value of Zn potential vs. Hg/HgO on discharge decreased as cycle number increased, even before discharge capacity faded.

The Zn electrode is associated with multiple failure mechanisms. Electrolyte depletion is due to a lack of hydroxide ions (OH^-), either because of OH^- consumption during battery discharge or an insufficient electrolyte reservoir. Pore plugging results from decreased electrode porosity associated with ZnO precipitation. Zn passivation occurs when precipitated ZnO films completely surround Zn. Shape change of the Zn electrode is a phenomenon in which Zn is not deposited uniformly during charge. The Zn electrode is also susceptible to gassing, in which the hydrogen (H_2) evolution reaction occurs during charge. Water is reduced to gaseous H_2, thus the flow of electrons into the electrode is not used to deposit metallic Zn, resulting in lower Zn plating efficiency.

In order to further understand underlying physical, chemical, and electrochemical behavior of the Zn electrode as well as improve energy density and cycle life, future experimental and theoretical work is proposed. The effect of additives on the cycle life of $Zn-MnO_2$ batteries will be studied. Additionally, electrode performance for Zn DOD values in the range of 15% to 30% Zn DOD will be investigated. A theoretical mathematical model of the Zn electrode will be formulated based on the governing equations of the system. Numerical solutions will be obtained for the spatial profiles of $Zn(OH)_4^{2-}$ concentration, OH^- concentration, current density, potential, and porosity at extended cycle number.

[1]Â Ingale, Nilesh D.; Gallaway, Joshua W.; Nyce, Michael; Couzis, Alexander; Banerjee, Sanjoy, â €œRechargeability and economic aspects of alkaline zinc-manganese dioxide cells for electrical storage and load leveling,â€ *Journal of Power Sources*, 276 (2015), 7-18.

[2] Wei, Xia; Desai, Divyaraj; Yadav, Gautam G.; Turney, Damn E.; Couzis, Alexander; Banerjee, Sanjoy. "Impact of anode substrates on electrodeposited zinc over cycling in zinc-anode rechargeable batteries,"Â Electrochimica Acta, 212 (2016), 603-613.

[3] Yadav, Gautam G.; Gallaway, Joshua W.; Turney, Damon E.; Nyce, Michael; Huang, Jinchao; Wei, Xia; Banerjee, Sanjoy. " Regenerable Cu-intercalated MnO_2 layered cathode for highly cyclable energy dense batteries," *Nature Communications*, 8 (2017) 14424.

[4] Turney, Damon; Gallaway, Joshua; Yadav, Gautam; Ramirez, Rodolfo; D'Ambrose, Michael; Kolhekar, Snehal; Nyce, Michael; Wei, Xia; Huang, Jinchao; Chen-Wiegart, Yu-chen; Wang, Jun; Banerjee, Sanjoy. "Rechargeable zinc alkaline anodes for long-cycle energy storage," *Chemistry of Materials*, under review.

487390 Degradation Modeling of Nickel-Rich Nickel-Cobalt-Manganese Oxide Cathode

Abstract:

As one of the promising cathode materials, nickel-rich nickel-cobalt-manganese oxide (Ni-rich NCM) has attracted major attention due to its large capacity and low manufacturing cost. However, its relatively poor cycle life hinders its realization. To achieve better cyclability, developing a comprehensive simulation model could be the most efficient way for preventing the degradation of Ni-rich NCM materials.

Several degradation behaviors have been revealed such as transition metal dissolution, electrolyte oxidation, salt decomposition and gas generation.[1,2] Especially for gas generation, the residual Li on the surface of the cathode such as LiOH and Li_2CO_3 is the main source of this phenomenon.[3]

In this study, we develop a electrochemical model coupled with the physics-based side-reaction for simulating the degradation of Ni-rich NCM cathode including gas generation. Most of known cathode degradation mechanisms are also considered in this study and implemented to the well-known Pseudo 2D electrochemical model.[2,4] To model the gas generation, oxygen activation and related role of residual Li are included by assuming CO_2 generation and oxygen activation reactions. Moreover, by constructing phase diagram, a critical level of delithiation of Ni-rich materials for oxygen generation is estimated. The corresponding equilibrium potential of oxygen activation reaction (U_{O2}) is estimated from the open circuit potentials of Ni-rich NCM materials, which is obtained from DFT calculation. The suggested model successfully reproduces the experimental charge/discharge voltage profiles within errors less than 5% up to 50 cycles. Especially, the large discrepancy of capacity between 1st and 2nd cycle observed from experiments is also captured from the simulation by only changing the critical U_{O2}. Based on the model, it will be discussed about the effect of residual Li and the positive correlation of metal dissolution and gas generation.

References

[1] K. Min, S.-W. Seo, Y.Y. Song, H.S. Lee, and E. Cho, Phys. Chem. Chem. Phys. **19**, 1762 (2017).

[2] X. Lin, J. Park, L. Liu, Y. Lee, A.M. Sastry, and W. Lu, J. Electro. Chem. Soc. **160**, A1701(2013).

[3] S. Hy, F. Felix, J.Rick, W.-N. Su, and B.J. Hwang, J. Am. Chem. Soc. **136**, 999(2014).

[4] M. Doyle, J. Newman, J. Electrochem. Soc. **143**, 1890(1996).

505271 Session Keynote - Beneficial Uses of Stranded Assets -- Emerging Opportunities

Abstract:

Stranded Assets: Industrialized economies have very high productivity of manufactured goods and agricultural products. This high productivity is associated with high level of wasted resources: for example waste heat from process plants, municipal solid waste going to landfills, agricultural biomass waste, sewage sludge, waste water and flare gas associated with oil and gas production. This wasted resources represent lost economic value, but also pose risk to environmental sustainability. These **Stranded Assets,** when converted to higher value co-products, represent multi-billion dollars market potential. States like California are leading the efforts and technology development to create Win-Win solution for these stranded assets as discussed next.

Emerging Opportunities: The wasted resources described above can be converted to higher value co-products such as hydrogen, methane, syn-gas, bio-oil, bio-char and chemicals. The conversion technologies include anaerobic digestion, gasification, pyrolysis, etc. among others. Efforts are underway to demonstrate these conversion technologies at commercial scale and validate them for economic viability and environmental sustainability. California is the first state to recognize the solid waste as a renewable energy source, extending it with policy support for financial incentives. One such example is: Hydrogen or Methane produced from human waste now qualifies for renewable fuel and low-carbon fuel credits. A Multi-Purpose Energy Station being developed by T2M Global Team offers a promising pathway for increased revenue from sale of higher value co-products, and thus a new opportunity of beneficial uses of the stranded assets. The innovative energy station provides H2 for fuel cell electric cars, charging for battery electric vehicles, methane for CNG vehicles, and grid support services to eliminate intermittency problems associated with solar and wind energy.

Policy Support Opportunities: Waste heat and flare gases deserve wasted resource recognition for their potential of economic value and environmental benefits. It may be appropriate to consider them as Tier 2 Renewables, similar to the precedence in CT, NY and NJ for clean power from natural gas fuel cells.

505134 Session Keynote: Low-Carbon Power Generation with Methane Fueled SOFCs

Abstract:

LOW-CARBON POWER GENERATION WITH METHANE FUELED SOFCs

Introduction

Natural gas fueled power generation systems, such as large scale thermal power plants and solid oxide fuel cells (SOFCs), are promising technologies with high electrical efficiencies of higher than 50% (LHV), and give high temperature exhaust heats[1]. While the CO_2 emission coefficient of the electrical power with a natural gas fuel will decrease with increasing the electrical efficiency, it remains to have non-negligible value even at the considerably high efficiency. Methane is a main component of natural gas. Recently thermal decomposition and catalytic thermal decomposition of methane have attracted attentions as advanced technologies producing COx-free hydrogen, which are simple ways to reduce the CO_2 emission caused by methane fuel[2].

Abstract

As a fuel of the SOFCs, methane is typically mixed with steam and reformed to hydrogen and carbon monoxide by steam-reforming reaction before supplied to SOFC stacks. The thermal energy needed for the endothermic steam-reforming reaction is supplied by the heat generated as a co-product accompanied by power generation from SOFCs. Methane fueled SOFCs have high electrical efficiencies, however, they accompany CO_2 emission of $0.198\ \eta^{-1}$ kg-CO_2 / kWh, where η is electrical efficiency. While the CO_2 emission coefficient decreases with increasing the electrical efficiency, the reduction of the coefficient has a lower limit of 0.198 kg-CO_2 / kWh by means of merely increasing the electrical efficiency.

We have conducted feasibility study of the innovative low-carbon methane-fueled power generation, in which methane was partly decomposed to form hydrogen and solid carbon before power generation, and a mixture of the generated hydrogen and undecomposed methane was used as a fuel. Figure 1 shows the schematic illustration of the system designed for the low-carbon power generation with the methane fueled SOFC. In the SOFC system, partially replacing the steam-reforming methane fuel by the decomposed methane fuel was found to result in a lower CO_2 emission coefficient of the electrical power generated in the system without a serious increase of the thermal energy needed for the fuel processing. The coefficient had strong dependence on the ratio of the decomposition to the steam-reforming.

The solid carbon deposited by the decomposition will be able to be utilized as carbon materials such as carbon black and carbon fiber, or source material of syngas which leads to liquid chemicals. Thus by using this system, carbon fixation in methane fueled power generation will be materialized through forming the co-products such as solid carbon, syngas, and/or liquid organic compounds.

Fig. 1. Schematics of the system used in this study,

Acknowledgement

This research is supported by The Japan Science and Technology Agency (JST) through its "Center of Innovation Science and Technology-based Radical Innovation and Entrepreneurship Program (COI Program).

References

1. Yoshio Matsuzaki, Yuya Tachikawa, Takaaki Somekawa, Toru Hatae, Hiroshige Matsumoto, Shunsuke Taniguchi, Kazunari Sasaki (2015)., "Effect of proton-conduction in electrolyte on electric efficiency of multi-stage solid oxide fuel cells", *Sci. Rep.* 5, 12640, pp 1-10

2. Hazzim F. Abbas, W.M.A. Wan Daud (2010), "Hydrogen production by methane decomposition: A review", *Int. J. Hydrog. Energy* 35 (3) pp. 1160-1190

PNNL-SA-130024

Pacific Northwest NATIONAL LABORATORY
Proudly Operated by Battelle Since 1965

Performance Modeling of Materials-Based Hydrogen Storage Systems for Automotive Applications

Kriston Brooks, Pacific Northwest National Laboratory
David Tamburello, Savannah River National Laboratory
Sam Sprik, National Renewable Energy Laboratory
Matthew Thornton, National Renewable Energy Laboratory

October 30, 2017

SRNL

NREL National Renewable Energy Laboratory

Pacific Northwest NATIONAL LABORATORY

U.S. DEPARTMENT OF ENERGY

Hydrogen Storage Engineering CENTER OF EXCELLENCE

November 21, 2017 | 1

Materials-Based Hydrogen Storage

Pacific Northwest NATIONAL LABORATORY
Proudly Operated by Battelle Since 1965

- Conventional H_2 Storage
  - Gaseous storage: 350 – 700 bar, ambient temperature
  - Liquid storage: 20K, near atmospheric pressure
  - Low volumetric and gravimetric density
- Materials-Based Hydrogen Storage
  - Adsorbents
    - Physisorption on high surface materials at 77K and ~ 100 bar
    - Activated carbons, MOF materials
  - Metal Hydrides
    - Reversible chemisorption: refuel with gaseous hydrogen
    - $NaAlH_4$, $TiCrMnH_2$, FeTi
  - Chemical Hydrides
    - Non-reversible chemisorption: return to central facility to regenerate
    - NH_3BH_3, AlH_3

November 21, 2017 | 2

Pacific Northwest
NATIONAL LABORATORY
Proudly Operated by Battelle Since 1965

Problem in Developing New Materials

- Challenge for Materials Developers
  - DOE Technical Targets are based on system properties not material properties
  - Material developers cannot quickly evaluate materials to determine if promising
    - Address engineering problems
    - Estimate storage system mass and volume
    - Determine system performance relative to drive cycles

DOE Technical Targets

- System Volume and Mass
- Vehicle Range
- Onboard Efficiency
- Vehicle Performance
  - Transient response
  - Refueling
  - Ambient temperature

November 21, 2017 | 3

Pacific Northwest
NATIONAL LABORATORY
Proudly Operated by Battelle Since 1965

DOE Hydrogen Storage Engineering Center of Excellence

- Multi-Lab/Industry project Under the Department of Energy (DOE) Fuel Cell Technologies Office
- Address the significant **engineering** challenges associated with developing materials-based hydrogen storage systems
- Enable fuel cell vehicles to meet customer expectations for driving range and performance for light duty vehicle applications
  - Design, build and evaluate **subscale prototype** systems.
  - Develop, validate, and disseminate new system, **engineering and design models**
  - Material developers use models to evaluate new materials.

NREL
Los Alamos NATIONAL LABORATORY
OSU Oregon State UNIVERSITY
HRI UQTR
GM
BASF
SRNL
United Technologies Research Center
MICHIGAN
Ford
JPL
HEXAGON LINCOLN
Hydrogen Storage Engineering CENTER OF EXCELLENCE

November 21, 2017 | 4

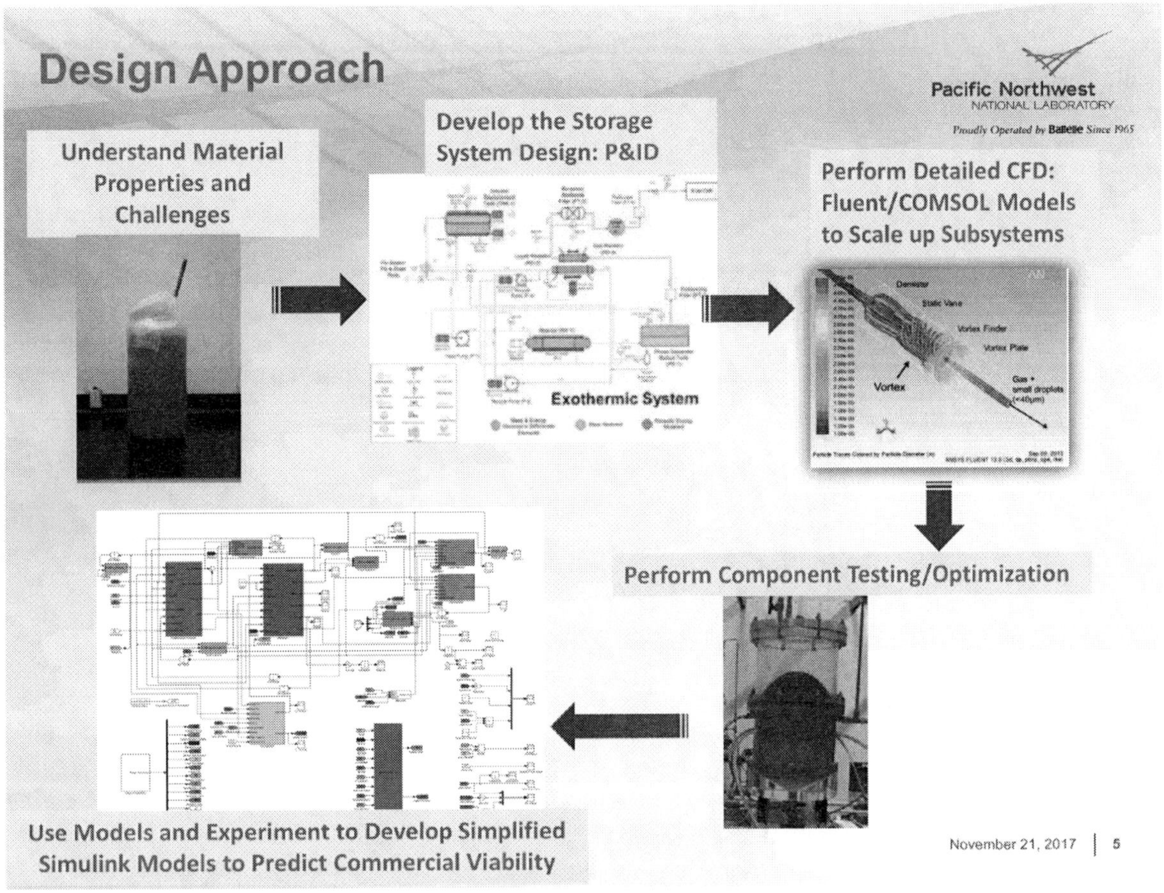
Design Approach
Pacific Northwest
NATIONAL LABORATORY
Proudly Operated by Battelle Since 1965
Understand Material Properties and Challenges
Develop the Storage System Design: P&ID
Exothermic System
Perform Detailed CFD: Fluent/COMSOL Models to Scale up Subsystems
Demister
Static Vane
Vortex Finder
Vortex Plate
Vortex
Perform Component Testing/Optimization
Use Models and Experiment to Develop Simplified Simulink Models to Predict Commercial Viability
November 21, 2017 | 5

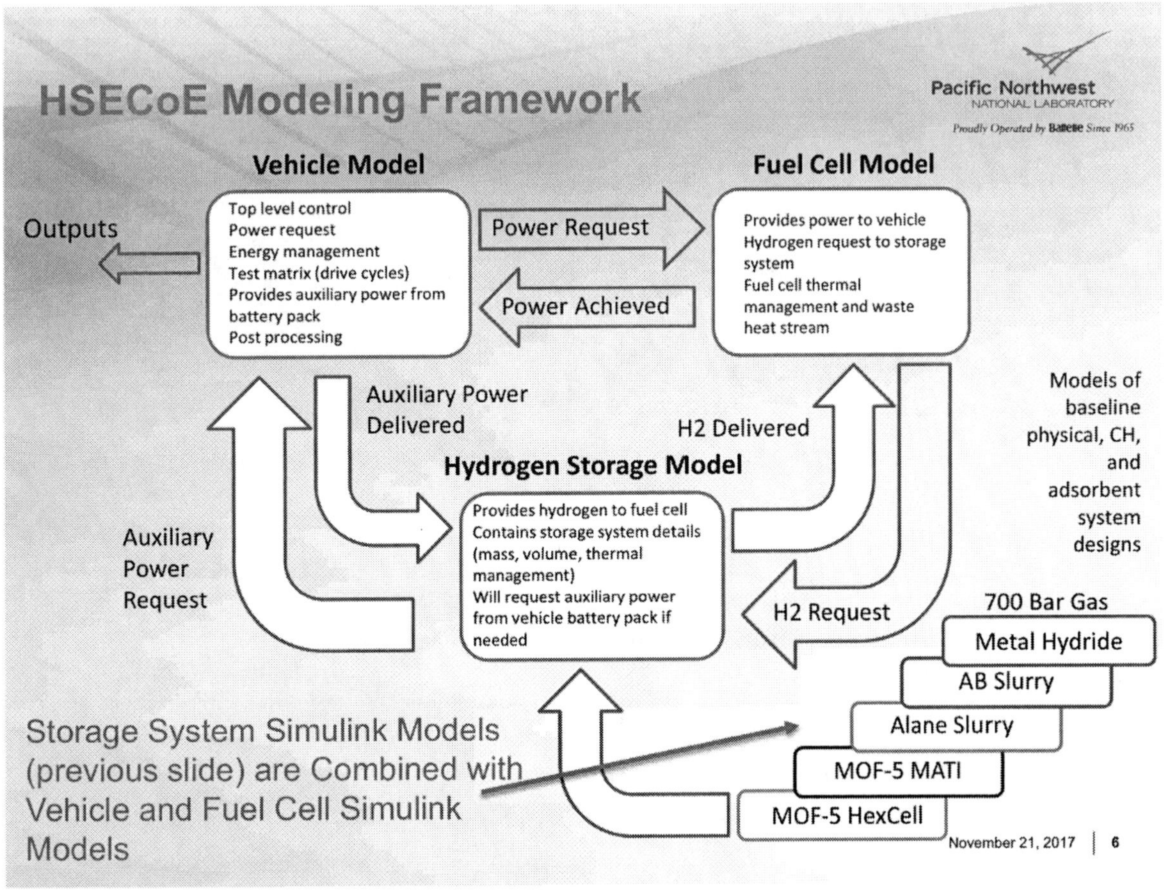
HSECoE Modeling Framework
Pacific Northwest
NATIONAL LABORATORY
Proudly Operated by Battelle Since 1965
Vehicle Model
Top level control
Power request
Energy management
Test matrix (drive cycles)
Provides auxiliary power from battery pack
Post processing
Outputs
Power Request
Power Achieved
Fuel Cell Model
Provides power to vehicle
Hydrogen request to storage system
Fuel cell thermal management and waste heat stream
Auxiliary Power Delivered
H2 Delivered
Hydrogen Storage Model
Provides hydrogen to fuel cell
Contains storage system details (mass, volume, thermal management)
Will request auxiliary power from vehicle battery pack if needed
Auxiliary Power Request
H2 Request
Models of baseline physical, CH, and adsorbent system designs
700 Bar Gas
Metal Hydride
AB Slurry
Alane Slurry
MOF-5 MATI
MOF-5 HexCell
Storage System Simulink Models (previous slide) are Combined with Vehicle and Fuel Cell Simulink Models
November 21, 2017 | 6

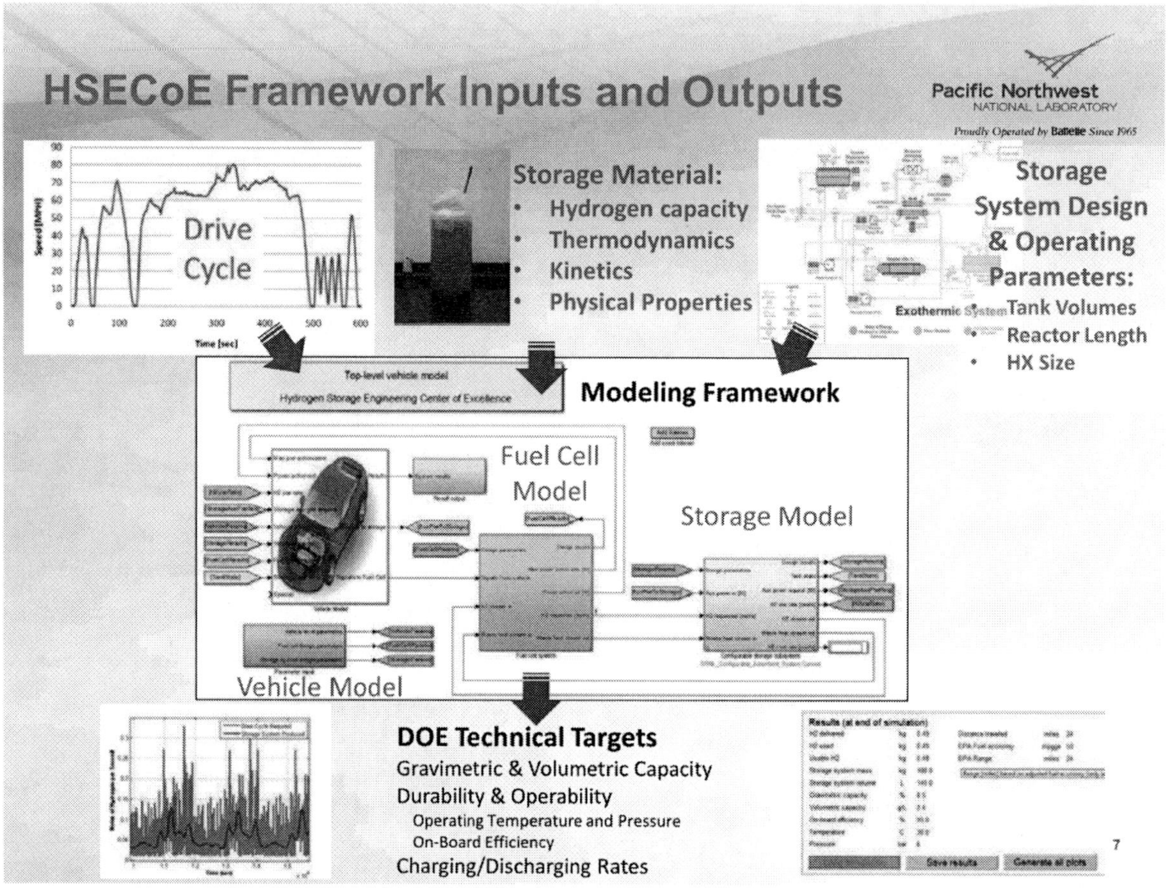
HSECoE Framework Inputs and Outputs
Pacific Northwest
NATIONAL LABORATORY
Proudly Operated by Battelle Since 1965
Drive Cycle
Speed [MPH]
Time [sec]
Storage Material:
• Hydrogen capacity
• Thermodynamics
• Kinetics
• Physical Properties
Storage System Design & Operating Parameters:
• Tank Volumes
• Reactor Length
• HX Size
Exothermic System
Top-level vehicle model
Hydrogen Storage Engineering Center of Excellence
Modeling Framework
Fuel Cell Model
Storage Model
Vehicle Model
DOE Technical Targets
Gravimetric & Volumetric Capacity
Durability & Operability
Operating Temperature and Pressure
On-Board Efficiency
Charging/Discharging Rates
Results (at end of simulation)
Save results
Generate all plots
7

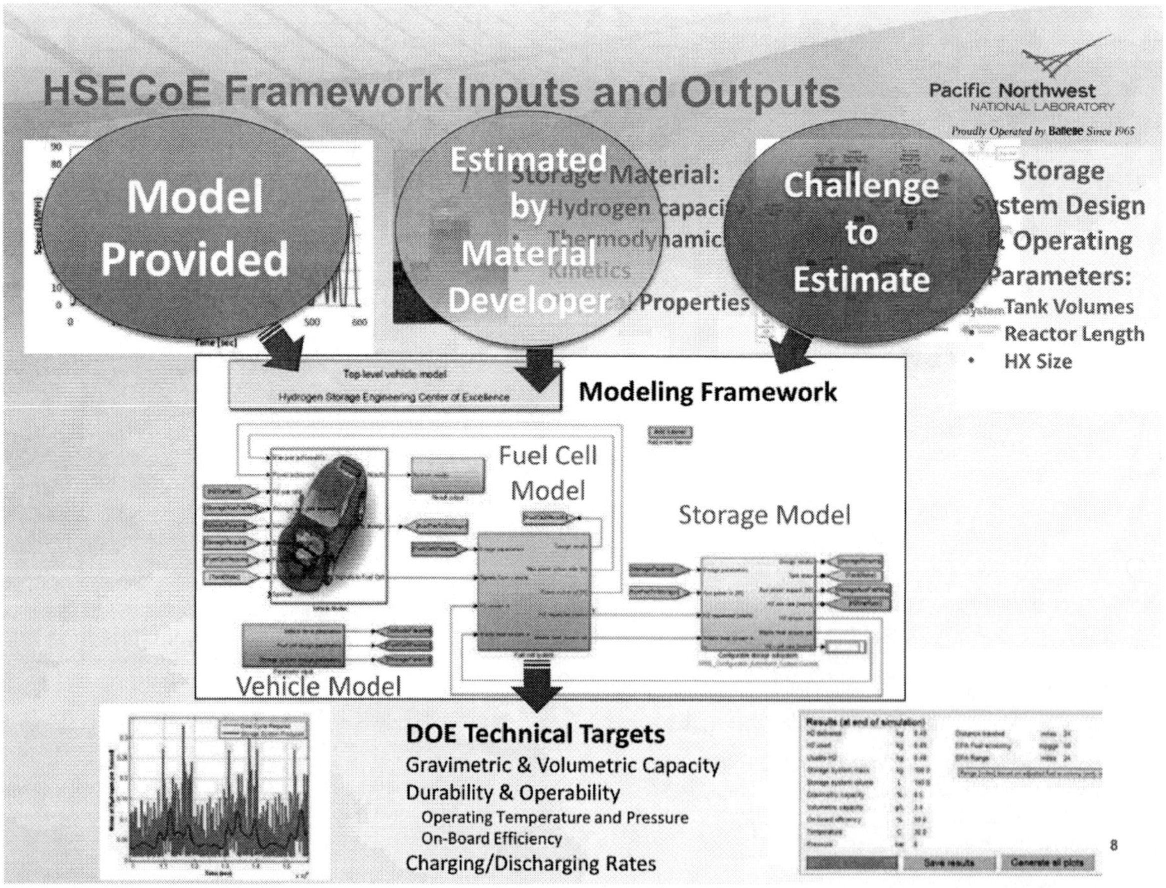
HSECoE Framework Inputs and Outputs
Pacific Northwest
NATIONAL LABORATORY
Proudly Operated by Battelle Since 1965
Model Provided
Estimated by Material Developer
Challenge to Estimate
Storage System Design & Operating Parameters:
Tank Volumes
Reactor Length
• HX Size
Top level vehicle model
Hydrogen Storage Engineering Center of Excellence
Modeling Framework
Fuel Cell Model
Storage Model
Vehicle Model
DOE Technical Targets
Gravimetric & Volumetric Capacity
Durability & Operability
Operating Temperature and Pressure
On-Board Efficiency
Charging/Discharging Rates
Results (at end of simulation)
Save results
Generate all plots
8

HSECoE Framework Inputs and Outputs

Pacific Northwest
NATIONAL LABORATORY
Proudly Operated by Battelle Since 1965

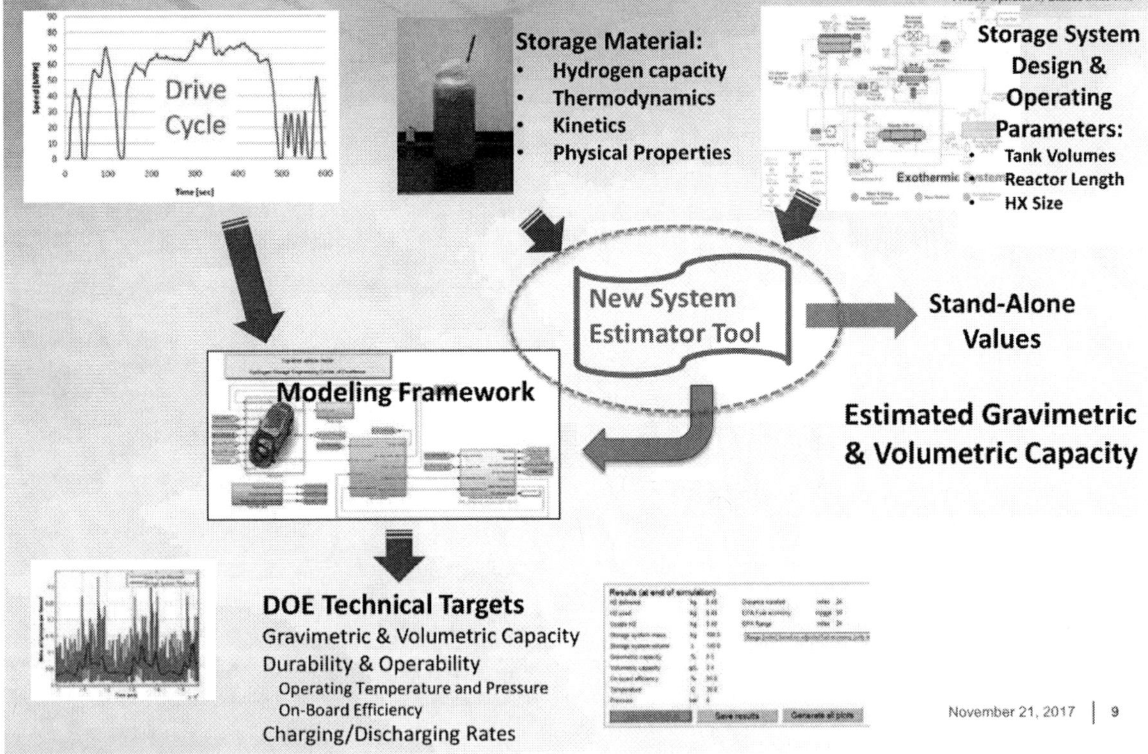

November 21, 2017 | 9

System Estimator Tool

Pacific Northwest
NATIONAL LABORATORY
Proudly Operated by Battelle Since 1965

- Developed for Adsorbents, Chemical Hydrides
  - Under development now for Metal Hydrides and Cryo-Compressed Systems
- Written in MatLAB
  - Can be included in the Framework
  - Can be run as a stand-alone executable file (no license required)
- User Inputs/Outputs
  - Microsoft Excel Table

November 21, 2017 | 10

Adsorbent System Design

Pacific Northwest NATIONAL LABORATORY

- Full tank pressure
  - 10 bar ≤ P_{tank} ≤ 100 bar
- Full tank temperature
  - 77 K ≤ T_{tank} ≤ 300 K
- Model was tuned toward a temperature increase of 80 K
  - Target usable hydrogen
  - 0.5 kg ≤ H2,usable ≤ 12 kg
- Type of Adsorbent / System
  - Adsorbent/System # 1 – Powder (0.18 g/cc)
    - HexCell heat exchanger
  - Adsorbent/System #2 – Compacted pucks (0.4 g/cc)
    - MATI isolated-fluid heat exchanger

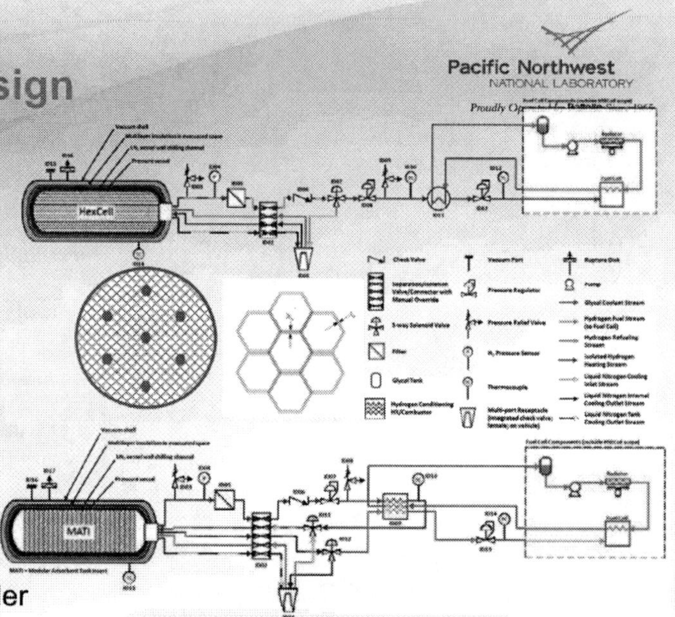

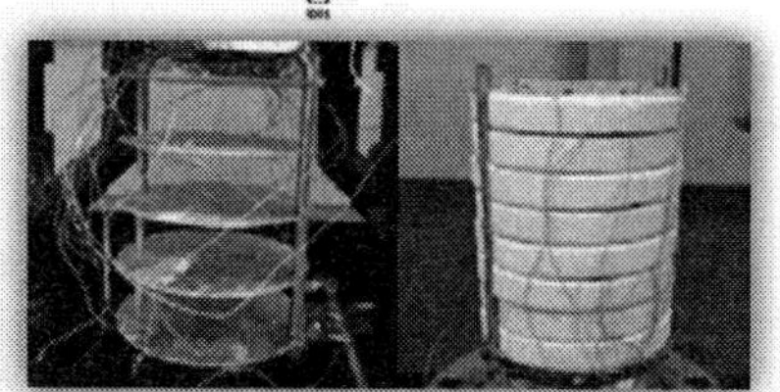

November 21, 2017 | 11

System Estimator Inputs for Adsorbents

Pacific Northwest NATIONAL LABORATORY

Proudly Operated by Battelle Since 1965

	A	B	C	D
1		Values	Units	Comments
2	Pi	1.00E+07	Pa	Initial/Full tank pressure
3	Pf	5.00E+05	Pa	Initial/Full tank temperature
4	Ti	80	K	Final/Empty tank pressure
5	Tf	160	K	Final/Empty tank temperature
6	H2usable	5.6	kg_{H2}	Target usable hydrogen
7	type_Ads	1		Type of adsorbent/HX: 1) Powder/HexCell, 2) Compact/MATI
8	alpha	2895.12802	J/mol_{H2}	D.A. Parameter -- Enthalpic contribution to the characeristic free energy of adsorption
9	beta	15.29117	$J/mol_{H2}/K$	D.A. Parameter -- Entropic contribution to the characeristic free energy of adsorption
10	m	2		D.A. Parameter -- Exponential constant for adsolute adsorption
11	nmax	96.43165	mol_{H2}/kg_{Ads}	D.A. Parameter -- Maximum H2 loading of the entire adsorption volume
12	P0	1.39E+09	Pa	D.A. Parameter -- Pseodo-saturation pressure (pressure of the gas phase)
13	rho_ads	130	kg_{Ads}/m^3	D.A. Parameter -- Bulk Density of the MOF-5
14	Va	0.00169712	m^3/kg_{Ads}	D.A. Parameter -- Adsorbed volume per mass of adsorbent
15	Vv	0.00725	m^3/kg_{Ads}	D.A. Parameter -- Void volume per mass of adsorbent
16	k	0.3	W/m/K	Thermal conductivity of the adsorbent
17	Cp	780	J/kg/K	Specific Heat of the adsorbent
18	Ads_Cost	11.8	$\$/kg_{Ads}$	Projected cost of the adsorbent per unit mass
19	th_ins	0.026	m	Pressure vessel insulation thickness
20	th_LN2	0	m	Pressure vessel LN2 chiller channel thickness (minimum value of 1/4" if preasent)
21	TType	4		Type of pressure vessel
22				0 = Aluminum Type 1
23				1 = 316 Stainless Steel Type 1
24				2 = Aluminum + CF Type 3
25				3 = SS + CF Type 3
26				4 = Plastic + CF Type 4

Operating Conditions (rows 2–5)

Adsorbent condition and Dubinin-Astakhov (D.A.) Parameters (rows 7–15)

Thermal properties (rows 16–17)

Pressure vessel design considerations (rows 19–26)

November 21, 2017 | 12

System Estimator Tool Results for Adsorbents

Pacific Northwest NATIONAL LABORATORY
Proudly Operated by Battelle Since 1965

	Gravimetric Capacity [g_{H2}/g_{sys}]	Volumetric Capacity [g_{H2}/L_{sys}]	Source
MOF-5 Powder [130 kg/m^3]	0.0338 g_{H2}/g_{sys}	18.6 g_{H2}/L_{sys}	HSECoE
MOF-5 Compact [406 kg/m^3]	0.0314 g_{H2}/g_{sys}	21.3 g_{H2}/L_{sys}	HSECoE
DUT-23 (Co) Powder* [200 kg/m^3]	0.0348 g_{H2}/g_{sys}	20.7 g_{H2}/L_{sys}	Ford / University of Michigan
IRMOF-20 Powder* [200 kg/m^3]	0.0341 g_{H2}/g_{sys}	20.3 g_{H2}/L_{sys}	Ford / University of Michigan
MOF-5 Powder [200 kg/m^3]	0.0332 g_{H2}/g_{sys}	19.6 g_{H2}/L_{sys}	HSECoE

***Special thanks to Ford and the University of Michigan for sharing their data**

System Design Assumptions:

- Operating Conditions: 80 K, 100 bar to 160 K, 5 bar
- 5.6 kg of usable hydrogen (~5.714 kg of actual hydrogen storage)
- Type 1 aluminum pressure vessel
- LN_2 pressure vessel chiller channel thickness of 9.525 mm
- Uniform insulation thickness of 23 mm, with a 2 mm outer aluminum shell

November 21, 2017 | 13

Chemical Hydride System Design

Pacific Northwest NATIONAL LABORATORY
Proudly Operated by Battelle Since 1965

- System pressure
  - 5 bar ≤ P_{system} ≤ 25 bar
- System temperature
  - 20°C ≤ T_{system} ≤ 300°C
- Model developed
  - exothermic
  - endothermic
  - liquid/slurry phase materials
- Target usable hydrogen:
  - 0.5 kg ≤ H2,usable ≤ 12 kg
- Volume Displacement Tank
- Thermolysis of reactants
- Phase Separation/Purification

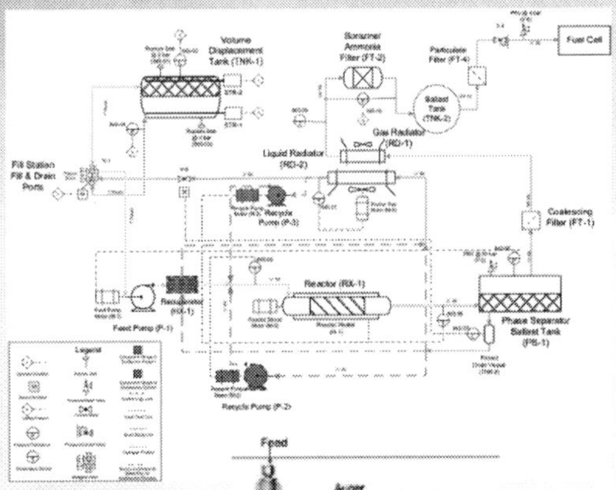

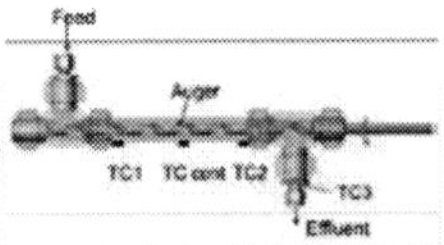

November 21, 2017 | 14

System Estimator Tool Inputs for Chemical Hydrides

Pacific Northwest
NATIONAL LABORATORY
Proudly Operated by Battelle Since 1965

	Values	Units	Comments
ExoEndo	1	--	Exothermic/Endotheric Flag (Exo = 1, Endo = 0)
Kinetic_Model	1	--	Kinetic Model Flag (Avrami Kinetics = 1, nth Order Kinetics = 2)
MW_CH	30.8	g/mol	molecular weight Chemical Hydrogen Material
slurry	1	--	Fluid Properties Flag (Slurry (1) or Liquid (0))
x_H2	0.152	--	Wt Fraction H2 in the CHS Material
n_rxn	1	--	Number of Reactions to Model (1 or 2)
DH_rxn_1	-17981	J/mol H2	Reaction Enthalpy Rxn 1 (negative=exothermic)
Beta1	2.355	mol H2/mol CH	Molar Ratio H2 maximum for CH material Rxn 1
A1	244	sec-1	Pre-exponential factor for Rxn 1
E1	29900	J/mol H2	Activation Energy for Rxn 1
n1	3.1	--	Exponent for Avrami or Reaction Order for Rxn 1
DH_rxn_2	0	J/mol H2	Reaction Enthalpy Rxn 2 (negative=exothermic)
Beta2	0	mol H2/mol CH	Molar Ratio H2 maximum for CH material Rxn 2
A2	0	sec-1	Pre-exponential factor for Rxn 2
E2	0	J/mol H2	Activation Energy for Rxn 2
n2	1	--	Exponent for Avrami or Reaction Order for Rxn 2
x_inert	0.5	--	Weight fraction inert with CHS Material to Slurry
Cp_CH	2694	J/kg/K	Heat Capacity CHS Material
Cp_i	1846	J/kg/K	Heat Capacity inert slurrying agent
Cp_p	774	J/kg/K	Heat Capacity CHS Material Product
rho_CH	780	kg/m3	Density CHS Material
rho_i	1000	kg/m3	Density inert slurrying agent
rho_P	1640	kg/m3	Density CHS Material Product
ppm_imp	500	ppm	Concentration of impurity 1
A_imp	0.1	g impurity/g adsorbent	Adsorbent maximum loading impurity 1
MW_imp	17	g/mol	molecular weight impurity 1
ppm_imp2	2000	ppm	Concentration of impurity 2
A_imp2	0.35	g impurity/g adsorbent	Adsorbent maximum loading impurity 2
MW_imp2	80.5	g/mol	molecular weight impurity 2
Useable H2	5.6	kg	Mass of usable hydrogen required
Power	40	kW	Average Hydrogen Storage H2 Production Required
Pset	25		Ballast Tank Pressure Initial Condition and Setpoint
Q_heater	8000		Reactor heater per length
Tmax	400	°C	Maximum acceptable reactor temperature

- Reaction Parameters, Weight Fraction H_2 (rows 2–18)
- Heat Capacity & Density (rows 19–24)
- Impurities (rows 25–30)
- Operating Conditions (rows 31–35)

November 21, 2017 | 15

System Estimator Results for Chemical Hydrides

Pacific Northwest
NATIONAL LABORATORY
Proudly Operated by Battelle Since 1965

Calculated System Parameter	Ammonia Borane	Alane	CBN*
Total System Mass (kg)	133	188	117
Total System Volume (L)	146	161	135
System Gravimetric Capacity (kg H_2/kg system)	0.042	0.029	0.048
System Volumetric Capacity (kg H_2/L system)	0.038	0.035	0.041
Reactor Length (m)	0.64	1.28	2.3
Ballast Tank Volume (L)	14	22	32
Mass Chemical Hydride (kg)	77	128	70
Fraction Chemical Hydride	0.5	0.5	1
Liquid Radiator Length (m) (3 tubes)	2.4	1.6	0.9
Gas Radiator Length (m) (1 tube)	1.2	1.0	1.4
Recuperator Length (m) (3 tubes)	0	3.22	0.8
Startup Temperature (°C)	178	202	279
Ballast Time (s)	75	117	176

*CBN:

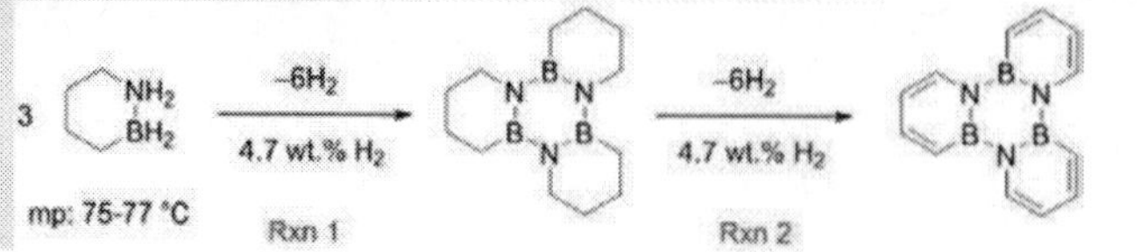

November 21, 2017 | 16

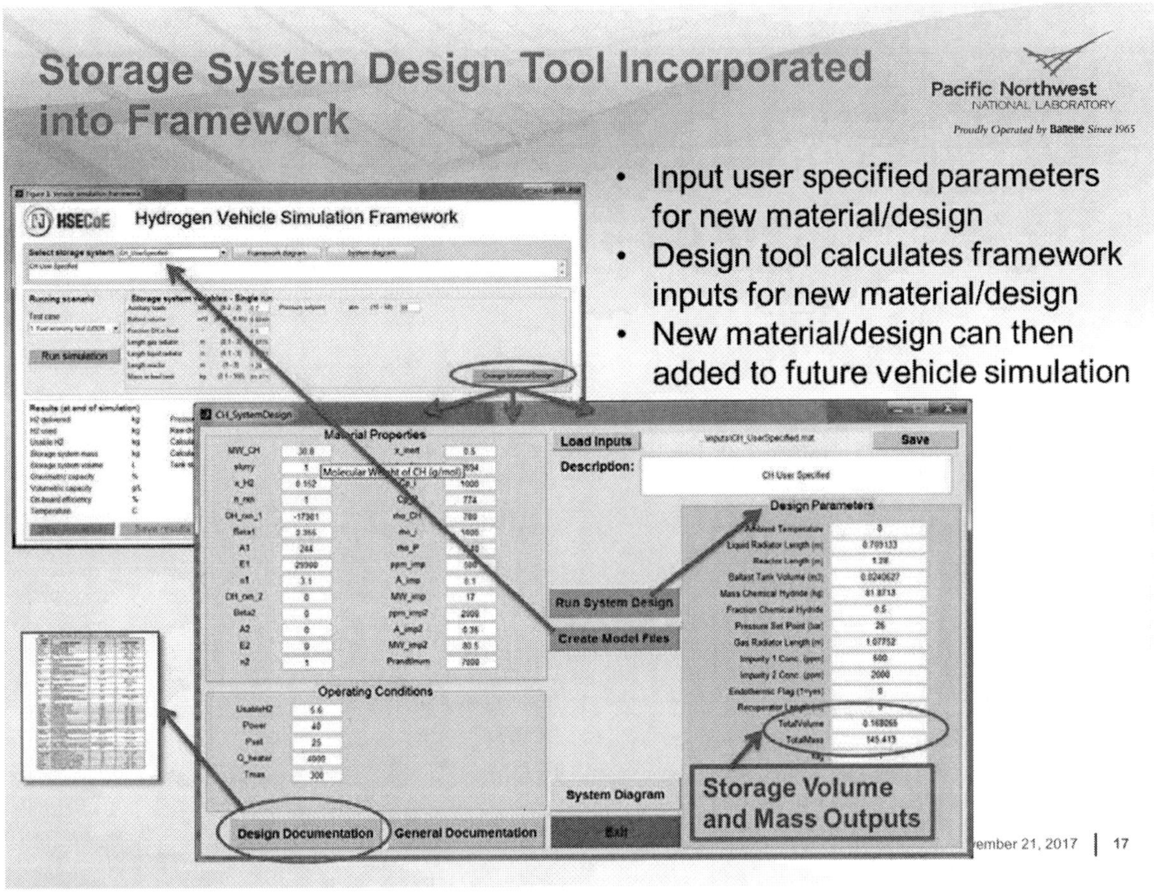
Storage System Design Tool Incorporated into Framework
Pacific Northwest
NATIONAL LABORATORY
Proudly Operated by Battelle Since 1965
Input user specified parameters for new material/design
Design tool calculates framework inputs for new material/design
New material/design can then added to future vehicle simulation
HSECoE
Hydrogen Vehicle Simulation Framework
Run simulation
Material Properties
Molecular Weight of CH (g/mol)
Operating Conditions
Load inputs
Save
Description:
CH User Specified
Design Parameters
Run System Design
Create Model Files
System Diagram
Exit
Design Documentation
General Documentation
Storage Volume and Mass Outputs
November 21, 2017
17

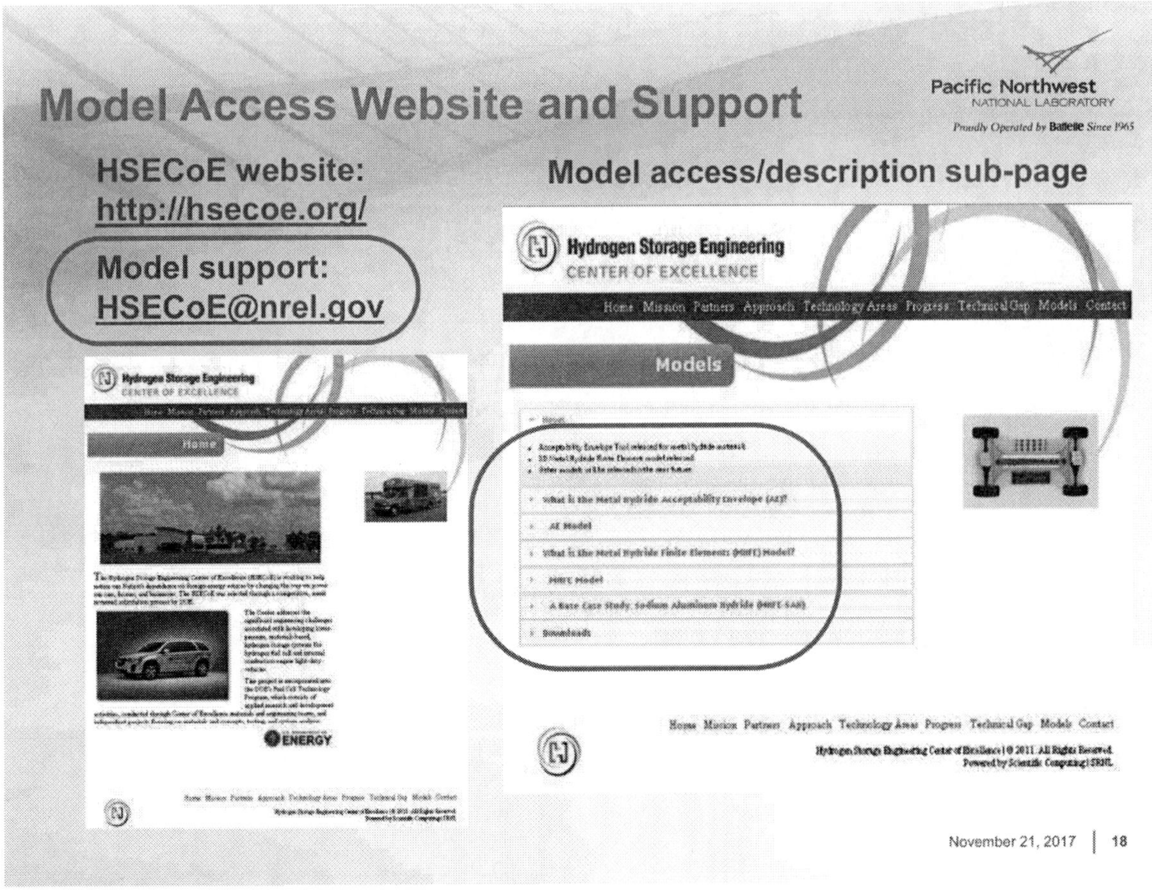
Model Access Website and Support
Pacific Northwest
NATIONAL LABORATORY
Proudly Operated by Battelle Since 1965
HSECoE website:
http://hsecoe.org/
Model support:
HSECoE@nrel.gov
Model access/description sub-page
Hydrogen Storage Engineering
CENTER OF EXCELLENCE
Home Mission Partners Approach Technology Areas Progress Technical Gap Models Contact
Models
Home
ENERGY
November 21, 2017
18

Storage System Models Available for Download

- Adsorbent Storage System Design Tool
- Chemical Hydride Storage System Design Tool
- Hydrogen Vehicle Simulation Framework
  - Chemical Hydride
    - Exothermic (AB)
    - Chemical Hydride Endothermic (Alane)
  - Adsorbent
    - Powder (HexCell)
    - Compacted Powder (MATI)
  - Metal Hydride (Generic hydride 30 kJ/mol)
  - Compressed Hydrogen (350 and 700 bar)

Conclusions

Pacific Northwest
NATIONAL LABORATORY
Proudly Operated by Battelle Since 1965

- Hydrogen storage using materials-based systems provide an alternative to conventional high pressure or liquid H_2 storage
- HSECoE developed the vehicle framework to evaluate hydrogen storage materials relative to the DOE technical targets
- Design tool allows material researchers to estimate
  - Parameters needed to run the framework
  - System mass and volume
- The framework and design tool have been developed for adsorbents and chemical hydrides.

Future Work

Pacific Northwest NATIONAL LABORATORY
Proudly Operated by Battelle Since 1965

- Develop design tools for cryo-compressed and metal hydrides
- Develop volume-based design to target specific vehicle volume limitations/designs
- Expand model to other vehicle classes (beyond light duty)
- Update Adsorbent model with Unilan (or the 2-state Langmuir) models in addition to the D-A model.
- Work with Material Based H_2 Storage Developers to Apply model to their materials

November 21, 2017 | 21

Questions?

Pacific Northwest NATIONAL LABORATORY
Proudly Operated by Battelle Since 1965

HSECoE Models on the WEB Team:

Matthew Thornton
David Tamburello
Kriston Brooks
Sam Sprik

With support from Bob Bowman and Mike Veenstra

Adsorption data provided by Ford, University of Michigan, and University of California Berkeley

Special thanks to the rest of the HSECoE, Jesse Adams, and Ned Stetson

November 21, 2017 | 22

501938 Immobilized Ferrite Nanoparticles for H_2 Generation *Via* Low Temperature Thermochemical Water-Splitting Process

Abstract:

H_2 in its cleaner form can be efficiently produced from the thermochemical water-splitting process. As the process is typically performed at higher temperatures of 900°-1600°C, significant energy input is needed for continuous H_2 production during multiple thermochemical cycles. At higher temperatures though grain growth and spalling of redox materials in thermal reactors often lead to decrease in the H_2 volume during multiple thermochemical cycles. In order to reduce the grain growth and achieve thermal stabilization, we attempted to synthesize novel immobilized ferrite nanoparticles using the sol-gel technique. In particular, highly porous ceramic supports were utilized to immobilize the ferrite nanoparticles. The morphologies of these immobilized materials were investigated by the scanning electron microscopy and X-ray diffraction. A number of ferrite nanoparticles were immobilized and their H_2 generation ability was compared at lower temperatures of 600°-900°
and H_2 volume generation results will be presented.

492691 Application of Liquid Organic Hydrogen Carrier in Mobility

Abstract:

A new, dense and safe form of transport of hydrogen is a Liquid Organic Hydrogen Carrier (LOHC). Meanwhile, the company Hydrogenious Technologies is selling loading and unloading equipment throughout the world.

A new challenge is to run the mobility on a CO2-free basis. One option is the battery car, but what still suffers from a very limited range and certain special drawbacks. An application to battery powered trucks for long distances is out of imagination.

So, the German government asked the author, to set up a plan, how these trucks can be operated with hydrogen in the form of LOHC. The lecture will describe the existing heavy duty hydrogen vehicles and the limits and advantages of this option.

501373 Super-Saturation of Nitrogen and Hydrogen in Transition Metals and Metal Hydrides/Nitrides

Abstract:

Super-saturation of metals with small gas molecules may occur when the amount of dissolved gas exceeds the theoretical maximum solubility at a given temperature and pressure. Since supersaturation effectively increases the amount of gas uptake, it could be useful in material storage and clean energy applications, such as in the case of hydrogen storage or carbon capture. One of the additional consequences of super-saturation is to induce a change in the physical properties of the bulk solid. Super-saturation may possibly enhance small gas diffusivity in metals, a factor that could improve separations in metallic membrane applications. A fundamental understanding of how super-saturation influences material properties can help guide engineers in the design of more effective materials.

The potential for super-saturation-induced property enhancement depends on two factors: (i) in what context is the desired property change considered desirable, and (ii) can super-saturation effectively induce this desired change. Other sub-related concerns involve the stability of this super-saturated state, the reversibility of the induced change, and the energy cost associated with establishing and maintaining the super-saturated state. In order to gain insight into the behavior of super-saturation in metals and related materials, density functional theory is used to investigate the fundamental mechanisms of super-saturation in both pure, alloy and defected (with vacancies) materials.

503670 Digital Alchemy for Assembly Engineeringc

Abstract:

Back in the day, Stuart Churchill was a pioneer in the use of computers for modeling chemical and thermal processes. Since that time, the role of computing in chemical engineering has become indispensable. In this talk, we present new computational methods we call "digital alchemy" that allow for predictive modeling of thermodynamic assembly, and permit the design of building blocks optimized for assembly into target structures with desired properties. In this approach, standard thermodynamic ensembles are generalized to include alchemical variables and their thermodynamic conjugates, allowing for rapid, real-time computational investigation of large, complex design spaces.

503671 Improved Algebraic, Numerical, and Graphical Representations in Fluid Mechanics

Abstract:

The concept of equivalents is applied to fully developed flows through straight channels, and two applications are introduced: the equivalents that produce equal rates of flow through channels of different cross-section, and the diminishment that occurs between the regimes of laminar and turbulent flow. The validity of the eddy diffusivity and the invalidity of the mixing-length as measures of the shear stress due to turbulence have perhaps never before been demonstrated so clearly. The possible choices of primary variables and the consequences are also examined. The objective of these explorations is an improved understanding of the basic fluid mechanics by teachers, students, and industrial practitioners.

503672 The Scaling of Turbulence Near the Wall and the Churchill Turbulent Flux Correlation: Insights with Lagrangian Simulations

Abstract:

The evaluation of mean velocity and mean temperature in wall-bounded turbulent flows has been based on the "classical" theory of turbulence and the law of the wall. The viscous wall parameters, namely the friction velocity, the friction temperature, and the kinematic viscosity of the fluid have been used to scale not only the mean values of velocity, but also other mean turbulent quantities close to the wall, including the Reynolds stresses and the turbulent velocity and temperature fluctuations. Stuart Churchill, as he questioned the validity of the Reynolds analogy between momentum and heat or mass transfer in his works [1], he realized that extending the scaling of momentum to the scaling of scalar quantities in turbulence was not a sound approach. He proposed a scaling of turbulence that depends on the turbulent momentum flux and the turbulent heat flux. In his AIChE Institute Lecture (in Miami, 1998), Churchill presented a summary of his work on this topic, and closed form solutions for the mean velocity and temperature [2]. The turbulent Prandtl number, the most significant parameter for the modeling of turbulent transport not only in theory but also in modern CFD commercial software packages, was also interpreted physically based on these ideas [3]. We have used direct numerical simulation results and results from Lagrangian simulations [4] for the transport of passive scalars in turbulent channel flows to explore Churchill's ideas on turbulent flow scaling and to develop an updated theory for the scaling of heat transport in turbulent flows [5-7]. His passion to remove empiricisms and approximations from the development of predictive correlations in turbulent transport has been an inspiration. In this talk, we will revisit the Churchill scaling in view of our simulation results, and will discuss the counterintuitive finding that wall turbulence can be used to separate rather than mix substances under special circumstances [8].

1. Churchill, S. W., 1996, "A Critique of Predictive and Correlative Models for Turbulent Flow and Convection," Ind. Eng. Chem. Res., 35, pp. 3122–3140.
2. Churchill, S.W., 2000, "Progress in the Thermal Sciences: AICHE Institute Lecture," AIChE Journal, 46(9), pp. 1704-1722.
3. Churchill, S. W., 2002, "A reinterpretation of the Turbulent Prandtl Number," Ind. Eng. Chem. Res., 41, pp. 6393–6401.
4. Papavassiliou, D.V, 2002, "Turbulent transport from continuous sources at the wall of a channel," Int. J. Heat Mass Transfer, 45(17), pp. 3571-3583.
5. Le, P.M., and Papavassiliou, D.V., 2008, "On the scaling of heat transfer using thermal flux gradients for fully developed turbulent channel and Couette flows," Int. Commun. Heat Mass Transfer, 35(4), pp. 404-412.
6. Srinivasan, C., and Papavassiliou, D. V., 2011, "Prediction of Turbulent Prandtl Number in Wall Flows With Lagrangian Simulations," Ind. Eng. Chem. Res., 50(15), pp. 8881–8891.
7. Srinivasan, C., and Papavassiliou, D.V., 2013, "Heat transfer scaling for wall bounded turbulent flows," Applied Mechanics Review 65(3), Art. 031002.
8. Nguyen, Q., Srinivasan, C., and Papavassiliou, D.V., 2015, "Flow induced separation in wall turbulence", Phys Rev E, 91, Art. 033019.

503674 Flow Boiling Using a Piranha Pin Fin Heat Sink

Abstract:

Flow Boiling using a Piranha Pin Fin Heat Sink

Corey Woodcock, Xiangfei Yu, Yoav Peles[1] and Joel Plawsky

Rensselaer Polytechnic Institute; [1]University of Central Florida

The *Piranha Pin Fin* (*PPF*) microdevice is a heat exchanger engineered to dissipate extreme thermal loads using either dielectric fluids such as HFE7000 or water. Fabrication of the device is CMOS compatible and the first generations of the device have been fashioned from silicon substrates. The PPF operates on the principle that heat transfer would be more efficient if heated fluids could be removed nearly as fast as they are created. This process would keep the driving force as high as possible. Thus the heat exchanger is built as two layers; a primary substrate that dissipates the bulk of the heat and shunts the heated fluids to a booster section that continues to heat the fluid and extends the total surface area of the device. Three generations of devices were fabricated. The first two generations were single-layer systems designed to help understand the operation of the primary layer and optimize the shape and location of the pin fins. The third generation was the first, two-layer system to be fabricated and tested.

Figure 1 Representative PPF array

Correlations for the heat transfer coefficient, h_{tp}, based on the Boiling Number and the Weber Number in the PPF design was developed and validated with Gen II. The discrepancy of heat transfer coefficient between predicted and measured values was within 20%. The ratio of two-phase to single-phase pressure drop was correlated with the Boiling Number (Bo) and the Jakob Number (Ja). The discrepancy of the two-phase multiplier between prediction and experimental measurements were within 12%.

The third generation *PPF* heat sinks were the first series to provide both a primary substrate populated with *PPF*s, as well as a booster substrate populated with solid micro pin fins. The substrates are micro-aligned and fusion-bonded to provide a full three dimensional (3D) architecture with bulk silicon properties. Experimental results from the Gen III *PPF* microdevices provide a proof of concept of the true *PPF* heat sink concept with fluid venting into multiple substrates. At low flowrates (G = 670 [kg/s/m^2]) the Gen III *PPF* heat sink has been shown to dissipate heat loads approaching 1.1 [kW/cm^2] at an exit quality of 95%, while at higher flowrates (G = 3,300 [kg/s/m^2]) the thermal limit of the heat sink remains unknown due to exceeding the available power supply at >1.4 [kW/cm^2].

503673 Transport Problems in the Spirit of Stuart Churchill for Teaching and Research at the University of Michigan

Abstract:

Stuart Churchillâ€™s solutions to important heat transport problems have greatly influence both pedagogy and research. The correlation of Churchill and Chu for the natural convection heat transfer to or from a vertical plate, over an astonishing range of Rayleigh numbers is based partly on a similarity solution of a type common in heat, mass, and momentum transfer for problems in chemical engineering. One such problem is asphaltene or wax transport and deposition onto the wall of an oil pipeline, which can be solved using similar methods, as will be discussed. In addition, such problems, and the wealth of their applications make for excellent motivations and examples in the teaching of graduate or undergraduate transport, as will be discussed in the talk.

504123 Session Keynote: Hydroxide Conducting Aromatic Polymer Membranes and Their Applications in Fuel Cells

Abstract:

Recently, anion exchange membrane (AEM) fuel cell has attracted attention as a clean energy technology device to convert chemical energy into electric energy. AEM fuel cells have several advantages, such as fast kinetics and non-precious metal catalysts.[1] However, AEMs that have desired properties (e.g. high hydroxide conductivity, thermochemical stability at high pH for durable operation, and convenient synthetic process) are not commercially available yet.[2-5] New material developments by designing of new polymer backbones, side chain functional groups and new anion conductive groups have been extensively investigated over the past decade.[6] Several research groups including us have reported that a long side chain between polymer backbone and quaternary ammonium (QA) end group is helpful for enhancing micro-phase separation based on thermochemical stable backbone.[7-9]

In this study, we will present the effects of polymer structures on the membrane morphology, properties and fuel cell performance. To better understand polymer structureâ€"membrane property relationships, aromatic polymers with different QA head groups, tether chain lengths, polymer backbone were prepared and their properties were evaluated. In addition to the fundamental study, how high performance AEMs can be produced in a kilo gram scale by cost effective processes will be presented.

References

1. Y. J. Wang, J. Qiao, R. Baker, J. Zhang, J., *Chem. Soc. Rev.* ***42***, 5768, (2013).

2. O. M. M. Page, S. D. Poynton, S. Murphy, A. Lien Ong, D. M. Hillman, C. A. Hancock, M. G. Hale, D. C. Apperley, J. R. Varcoe, *RSC Adv. 3*, 579 (2013).

3. K. J. Noonan, K. M. Hugar, H. A. Kostalik, E. B. Lobkovsky, H. D. Abruna, G. W. Coates, *J. Amer. Chem. Soc.* ***134***, 18161 (2012).

4. J. Ran, L. Wu, T. Xu, *Polym. Chem. 4*, 4612 (2013).

5. C. Fujimoto, D.-S. Kim, M. Hibbs, D. Wrobleski, Y. S. Kim, *J. Membr. Sci.* ***423â^'424***, 438 (2012).

6. A. D. Mohanty, C. Y. Ryu, Y. S. Kim, C. Bae, *Macromolecules* ***48***, 7085 (2015).

7. H.-S. Dang, P. Jannasch, *Macromolecules* ***48***, 5742 (2015).

8. W.-H. Lee, Y. S. Kim, C. Bae, *ACS Macro Lett.* **4**, 814 (2015).

9. W.-H. Lee, A.D. Mohanty, C. Bae, *ACS Macro Lett.* **4**, 453 (2015).

504854 Session Keynote: Gas-Transport Limitations in Electrochemical Energy-Conversion Technologies

Abstract:

The issue of transport phenomena is omnipresent in various electrochemical energy-conversion technologies. There is a need to understand and optimize the various means for products and reactants, whether gas, liquid, or ions to get to and away from the reaction sites. Such transport often results in concentration polarization and ohmic effects in various electrochemical cells. Mathematical modeling and advanced diagnostics are ideally suited to elucidate and explore the impacts of transport phenomena on device performance.

In this talk, we will discuss current research questions and issues related to transport phenomena in various electrochemical energy-conversion technologies including solar fuel generators, polymer-electrolyte fuel cells, and redox flow batteries. These various technologies share very similar architectures of typically ion-conducting polymer membranes sandwiched between thin catalyst layers that are porous electrodes. Key in many of these technologies is the need to get the reactant gas to the reaction site within the porous electrodes. In this respect, two main technologies will be explored, namely, vapor-fed solar-fuel generators and polymer-electrolyte fuel cells. If time permits, exploration of ion transport and selectivity in redox flow batteries will be discussed.

For polymer-electrolyte fuel cells, the drive to decrease the amount of active catalyst has resulted in additional, unexpected mass-transport losses to the reaction site. It is believed that these losses are due to gas transport through the ionomer film covering the particles. In this talk, we will discuss measurement at limiting current of these resistances as well as the associated ionomer film morphology to help shed the light on this issue.

For solar-fuel generators, the need to remove the inherent carbon-dioxide transport and solubility issues drives the exploration of gas-diffusion electrodes. In this talk, we will explore how the various transport limitations come into play and how the structure of electrode dictates product selection and yield.

Acknowledgement

This work was funded by the Department of Energy under contract number DE-AC02-05CH11231.

499577 Experimental Measurements of Water Transport in Proton Exchange Membrane Fuel Cells Via in-Situ Performance Testing and Ex-Situ Synchrotron X-Ray Radiography

Abstract:

Proton Exchange Membrane fuel cells are an electrochemical energy conversion device offering the potential for high power density and no harmful emissions at the point of use. As hydrogen and oxygen react, the byproducts are electricity, heat, and water. Managing that product water is essential to maintain high performance and efficiency. Excessive water in the flow channels or porous electrode materials can increase the system pressure drop and lower power output. An impediment to studying water in the fuel cell itself is that the fuel cell materials are opaque, hindering direct optical visualization. The current study addresses this issue via novel performance testing methods of an operating (in-situ) fuel cell and Synchrotron X-ray radiography testing of the porous materials in a non-operating (ex-situ) setup. In the in-situ testing, the fuel cell is initially operated at typical operating conditions (75°C, fully humidified inlet gases, inlet pressure at 206.9 kPag, and an operating current density of 1000 mA/cm^2). To initiate the testing, the anode (hydrogen) flow is switched to a dry inlet, creating a concentration gradient that favors water transport to the anode. At two minute intervals, the anode flow rate is increased in stepwise fashion over a range of stoichiometries from 1.5-10, further increasing the water removal capacity of the anode at each step. At each interval, the cell voltage, the anode pressure drop, and the cathode pressure drop are measured. Importantly, a calibration curve of the anode pressure drop at several relative humidity values is used to determine the relative humidity of the anode throughout this operation. This relative humidity measurement, and knowing the temperature and pressure, thus gives a quantitative estimate of the amount of water transported to the anode at each flow interval. Knowing this value, estimates can be made related to the magnitude of water accumulating in the porous layer and the amount accumulating in the cathode (air) flow channels. Tests showing greater water removal also show substantial increases in voltage during the test, indicative of the porous layer being saturated. To corroborate the data related to saturation of the porous layer, Synchrotron X-ray radiography at the Canadian Light Source is used with an acrylic setup that contains commercially available graphite porous layers. Water is injected through a plenum below the porous layer while heated air flows through channels above, mimicking fuel cell operation. The acrylic limits attenuation, providing a more accurate measurement of water accumulated in the porous layer. By considering a range of flow rates and temperatures, the removal rate of water from the porous layer is quantified through 2D radiographs and 3D computed tomography scans.

509054 Solve this! Fundamental Approach to Problem Solving in Industrial Processes I (Invited Talks)

Abstract:

Solving complex plant problems is an everyday exercise that affects all projects and budgets, and it is very rarely introduced to students at any level. In fact, by the time young engineers are few years into working in industry, most have lost the engineering skills that they developed in school. We hope to encourage future engineers to retain those skills and learn the importance of plant problem-solving.

505001 Session Keynote: Economics of Large Scale Hydrogen Generation: Can H_2 Compete with Batteries?

Abstract:

Interest in hydrogen, always vulnerable to the changing winds of government and industrial interest once again finds itself at a peak. Electrolysis of water, when coupled with renewable energy is generally accepted as the only carbon-free method of hydrogen production. Alkaline hydrolysis which has long been the work-horse of large-scale electrolysis is generally being replaced by PEM (polymer electrolyte membrane) due to the latter's higher efficiency, and ability to provide higher and differential pressure. Subsequently there are more and more providers of large-scale PEM systems. Despite this there is little agreement on what the cost of that hydrogen will be. Charlie Freese, executive director- global fuel cell activities at General Motors recently gave a target of $0.57/kg.[1] though current suppliers of hydrogen (from centralized reformation) are charging $16.00/kg, using natural gas, which is currently the cheapest method to supply hydrogen.

The difficulties in generating, developing and supplying high pressure (700 bar) hydrogen coupled with rapidly decreasing battery costs has made batteries the current preferred choice for both electric vehicles and grid level storage. Can hydrogen compete? This talk will look at the underlying costs of hydrogen from electrolysis. It will also look at the opportunities and the technical challenges of lowering both capital and operating expenditures. Specifically, the role of noble metals, often the focus of PEM research funding, will show that they do not contribute significantly to the cost of electrolysis at current usage. Finally much of the assumptions of low-cost hydrogen from electrolysis is based on cheap, or even free energy from excess renewables. Taking advantage of these low-cost intermittent sources can greatly reduce the utilization of the electrolysis capital equipment however, greatly increasing capital costs.

[1] Amend, James M. "Fifty-Cent Fuel Ahead, If Stars Align" WardsAuto June 2016. http://wardsauto.com/engines/fifty-cent-fuel-ahead-if-stars-align

505100 Session Keynote: Utilizing Proton Exchange Membrane (PEM) Electrolyzers as a Robust Platform for Emerging Electrochemical Technologies

Abstract:

Proton OnSite, a world-leader in hydrogen gas solutions, manufactures proton exchange membrane (PEM)-based electrolyzers which use pure water and electricity to generate hydrogen and oxygen gas. PEM-electrolysis is highly scalable from watts to megawatts, is reliable and durable with low maintenance requirements. This technology can be easily integrated with renewable energy, making hydrogen generation emission-free. This technology is rapidly gaining international interest for energy storage applications due to the flexibility and deployability it provides. Key advantages include the ability to interconnect various infrastructures such as renewable electricity, transportation, and biogas upgrading, as well as the ability to address very long timescales, such as seasonal variations in solar intensity.

Solid membrane systems for hydrogen generation are well established commercially, and serve as an excellent basis for emerging electrochemical technologies. Proton OnSite has flexed their already existing platforms, as well as their extensive stack design and system expertise to develop exciting new ways to utilize and harness energy. They have successfully worked with many partners in academia and industry to advance materials and systems in the areas of electrochemical ammonia generation, microbial fuel cells for water remediation, electrochemical pumping and flow battery technology. With further development, a variety of electrochemical solutions could create an incredibly efficient and low-polluting energy landscape.

Proceedings of the 2017 AIChE Annual Meeting
October 29- November 3, 2017, Minneapolis, MN

Demonstration of Hydrogen Production by Multi-typed SOEC system

Naomi Tsuchiya[1], Hisao Ohmura[1], Shohei Kanamura[1], Masato Yoshino[2], Seiji Fujiwara[3], Tsuneji Kameda[4], Masahiko Yamada[5], Kazuya Yamada[1]

(1)Technology Management Div., Toshiba Energy Systems & Solutions Corporation, Kawasaki, Japan,

(2)Power & Industrial Systems Research & Development Center, Toshiba Corporation, Yokohama, Japan,

(3)Isogo Nuclear Engineering Center, Toshiba Energy Systems & Solutions Corporation, Yokohama, Japan,

(4)Power & Industrial Systems Research & Development Center, Toshiba Corporation, Fuchu, Japan,

(5)New Energy Solutions Project Team, Toshiba Energy Systems & Solutions Corporation, Kawasaki, Japan

ABSTRACT

To establish a hydrogen production technology based on renewable energy, we are developing a hydrogen production system based on high-temperature steam electrolysis using solid-oxide electrolysis cells (SOECs). We have already developed a cell and cell-stack that produce hydrogen more efficiently than water electrolysis. For the mass production of hydrogen, it is necessary to connect multi-typed cell-stacks and to increase the electrolysis capacity. Moreover, it is essential to use heat effectively to reduce the energy consumption of the SOEC system since the SOEC operating temperature range is from 600 to 800 °C. In this study, we evaluated the following: 1) the effect of scaling-up by constructing multi-typed cell-stacks, 2) the effect of exhaust heat recovery on reducing energy consumption, 3) the influence of fluctuations in electrolytic power on a hydrogen production system in which hydrogen production is performed using renewable energy.

I. INTRODUCTION

One hydrogen production technology that is required in order to realize a hydrogen society is hydrogen production based on electrolysis of water. High-temperature steam electrolysis using solid-oxide electrolysis cells (SOEC) can produce hydrogen at a lower electrolytic voltage as compared with alkaline-water electrolysis or polyelectrolyte film (PEM) water electrolysis. Moreover, since the material involved is only water, this process does not directly generate carbon dioxide, which is considered to be one of the causes of global warming.

Therefore, a hydrogen feed system that does not discharge CO_2 can be established by producing hydrogen using photovoltaic generation, wind power generation, etc.

Toshiba is engaged in research and development with the aim of realizing technologies for a next-generation hydrogen society in which carbon dioxide is not emitted.

So far, a cell-stack, which is a key component of an SOEC hydrogen production system, has been developed, allowing hydrogen production to be carried out more efficiently than water electrolysis. However, in order to produce hydrogen in large quantities, it is necessary to connect two or more cell-stacks to increase the electrolysis capacity from the limited amount of hydrogen production possible with a single cell-stack.

Moreover, in a hydrogen production system, considering that the energy consumed by heating components other than the cell-stack is large in order that the SOEC may operate at a high temperature of 600–800 °C, the merits of an SOEC hydrogen production system will be lost. Therefore, it is essential to use heat effectively in order to reduce the energy consumption of the whole system.

In this research, we evaluated the effect of scaling-up by employing multi-typed cell-stacks, and we evaluated the reduction in energy consumption made possible by exhaust heat recovery.

Moreover, the influence of changes in the electrolysis energy on the hydrogen production was evaluated by assuming that hydrogen production was performed using renewable energy.

II. Principle of High-temperature Steam Electrolysis by SOEC

Figure 1 shows the basic electrolysis scheme. In the case of SOEC, the basic structure and materials have much in common with Solid Oxide Fuel Cell (SOFC) technology, but the electrolysis cell works in reverse compared with that in SOFC. During electrolysis, hydrogen is produced from supplied steam at the cathode (hydrogen electrode), as described in equation (1), and oxygen is released at the anode (oxygen electrode), as described in equation (2).

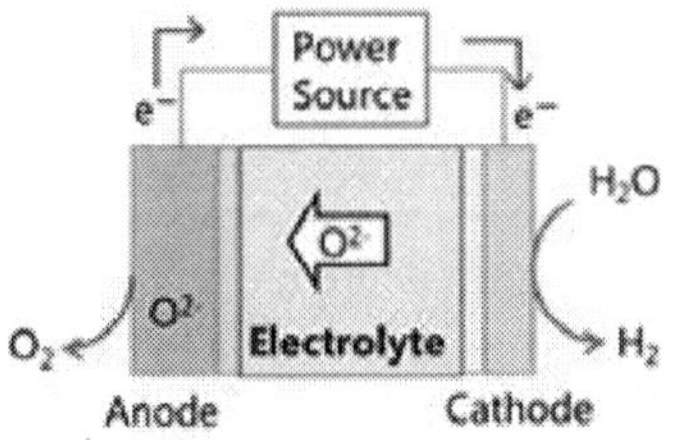

Figure 1 Scheme of electrolysis cell

$$H_2O + e^- \rightarrow H_2 + 1/2\ O_2^- \qquad (1)$$
$$1/2\ O_2^- \rightarrow 1/2\ O_2 + e^- \qquad (2)$$

The enthalpy change by water electrolysis can be described by equation (3)

$$\Delta H = \Delta G\ +\ T\Delta S \quad (3)$$

Here, ΔG is equivalent to electrical energy, and $T\Delta S$ is equivalent to thermal energy. A feature of this high-temperature steam electrolysis is that part of the energy required for electrolysis can be provided using heat. Thus, as shown in Figure 2, high-temperature steam electrolysis by SOEC at an operating temperature above 600 °C basically requires less electrical energy compared to other electrolysis methods operating at lower temperatures.

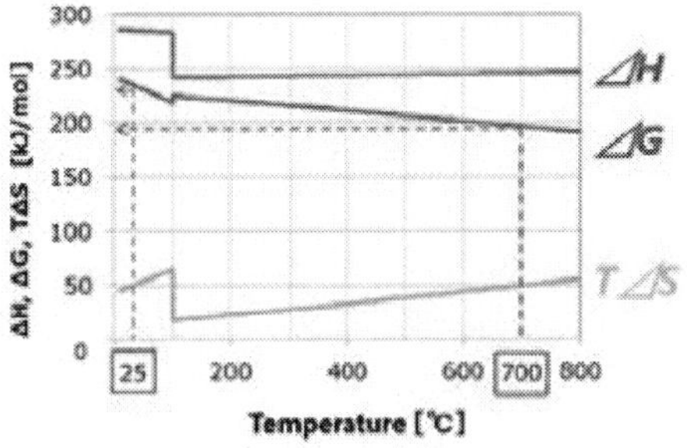

Figure 2 Required energy for electrolysis

III. Test Plant

The basic configuration of a multi-typed cell-stack pilot test plant is shown in Figure 3. This test plant consisted of the following components: an electric furnace (multi-typed cell-stack container), a steam generator, a steam supply line, a produced hydrogen exhaust line, an air supply line for diluting and exhausting by-product oxygen, and an oxygen-rich air exhaust line. The steam supply line and the produced hydrogen exhaust line have a heat exchanger, and the air supply line and the oxygen-rich air exhaust line also have a heat exchanger.

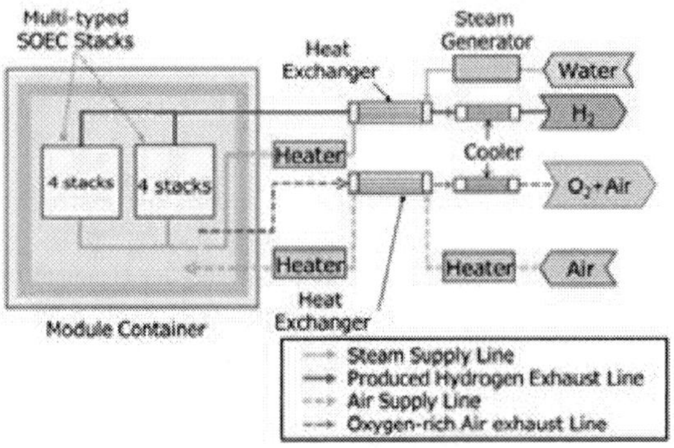

Figure 3 Basic configuration of multi-typed cell-stack pilot test plant

The thermal energy of the high-temperature fluid exhausted from each multi-typed cell-stack container is given to the low-temperature supply gas via the heat exchangers. These heat exchangers were designed so as to minimize the energy that heats the supply gas.

Furthermore, in consideration of the pressure loss in the cell-stacks, orifices were provided in the steam supply line so that gas was uniformly supplied to each cell-stack.

A schematic diagram of the gas line in a multi-stack container and a current line is shown in Figure 4. In this container, 8 cell-stacks are provided, composed of 4 cell-stacks connected in series in 2 lines. Each line is connected to a DC power supply unit, so the currents can be individually controlled.

Finally, the hydrogen generated by electrolysis is collected in one line and is exhausted out of the multi-stack container.

The product oxygen is exhausted to the outside of cell-stacks and remains in the container. Therefore, this test plant was designed to supply air into the container and exhaust product oxygen with air.

The total electrolysis power was about 10 kW.

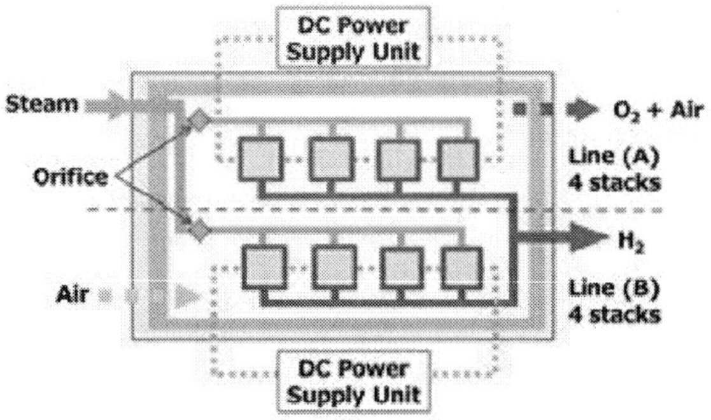

Figure 4 Schematic diagram of the gas line in the multi-stack container and current lines.

IV. Hydrogen Production Test

The specifications of the hydrogen production test plant are shown in Table 1. In this test, in order to control oxidization of the electrode, hydrogen was added together with steam. The temperature in the module container was 700 °C, and the steam concentration of the feed gas was from 90% to 95%.

Table 1 Specifications of test plant.

Item		Value
Cell-stack		4 stacks×2 lines (about 10kW)
Temperature in Module container		700℃
Feed gas	Cathode	Steam 3.78 kg/h H_2 0.25~0.50 Nm³/h ※ Steam conc. :90~95%
	Anode	Air 5.6 Nm³/h

First, the current-voltage (I-V) characteristic of the cell-stacks was measured by applying a current of 0 to 21 A to each cell-stack. When measuring the I-V properties, after setting the current as a test condition, it was checked that the voltage and temperature of the cell stack had become constant.

The I-V properties of the cell-stack when formed into the multi-stack and a single cell-stack are shown in Figure. 5. This test was carried out by supplying a feed steam concentration of 90 %. As shown in Figure 5, there was no large difference in the cell voltages of the 8 cell-stacks, and the I-V properties of the 8 stacks were almost the same as that of a single cell-stack. Therefore, it was confirmed that steam was evenly supplied to each cell-stack.

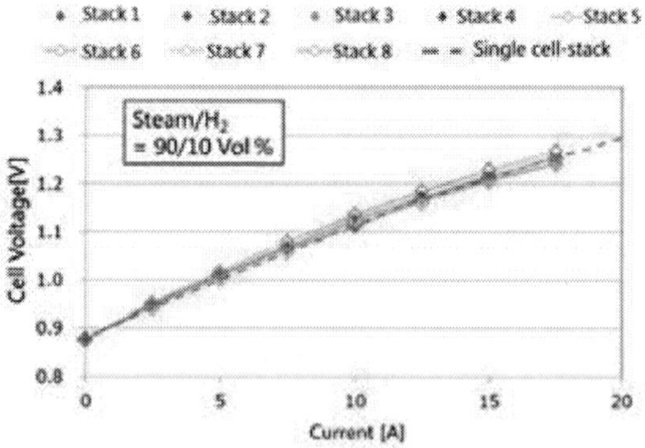

Figure 5 I-V properties of multi-typed cell-stacks and single cell-stack.

When acquiring the I-V characteristic, the hydrogen production rate is also measured and compared with the theoretical hydrogen production rate calculated from the current applied to the multi-typed cell-stacks.

A comparison of the measured value and a theoretical value is shown in Figure 6. The hydrogen production rate in the test was almost the same as the theoretical value. When a current of 20 A (current value near the electrolysis heat neutral point) was applied to each cell-stack, the hydrogen production rate was 3.14 Nm^3/h. This is eight-times greater than the hydrogen production rate (0.39 Nm^3/h) when a current of 20 A was applied to a single cell-stack. Therefore, the feasibility of scaling-up by employing multi-typed cell-stacks is confirmed.

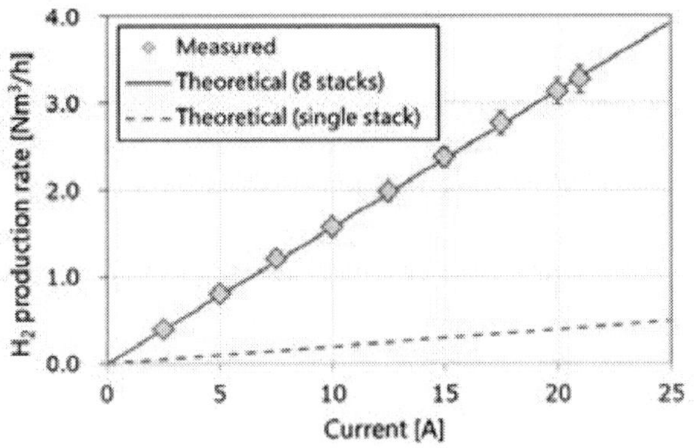

Figure 6 The relationship between hydrogen production rate and supply current.

V. Reduction in Energy Consumption by Exhaust Heat Recovery

It is necessary to take into consideration three points: the heat loss due to energy required for electrolysis, energy required for heating the supply gas, and heat dissipation in the SOEC hydrogen production system.

In order to reduce the energy of the SOEC system and to reduce the energy required for producing hydrogen at a rate of 1 Nm^3/h, it is necessary to reduce the energy required for processes other than electrolysis.

Figure 7 shows the process flow in the case where heat exchangers are not included, and Figure 8 shows the process flow we designed. When there was no heat exchanger, it was assumed that water, steam, and air supplied to the multi-typed SOEC container were heated from 25 to 700 °C by the steam generator and heaters. Heat dissipation of the container was estimated from the size of the container and the thickness of thermal insulation etc.

In this case, we estimated the heat dissipation of the container only because its surface area was the largest.

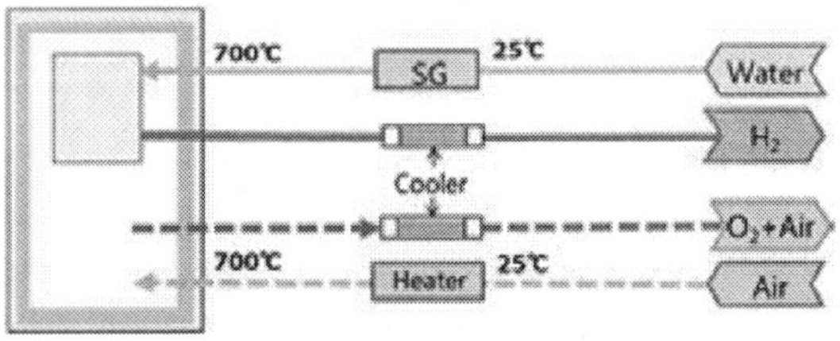

Figure 7 Process flow in case where heat exchangers are not included.

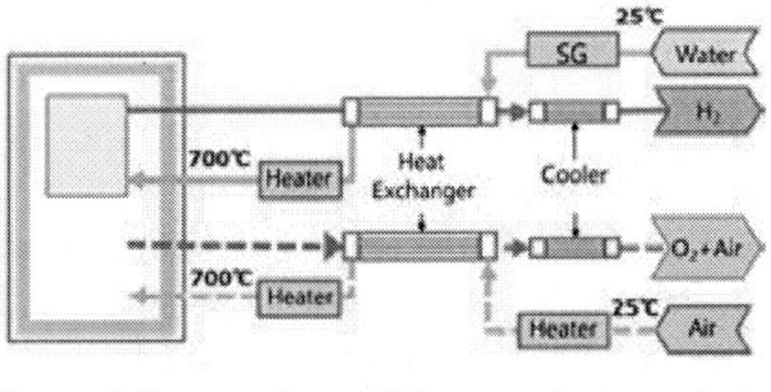

Figure 8 Process flow of the test plant we designed.

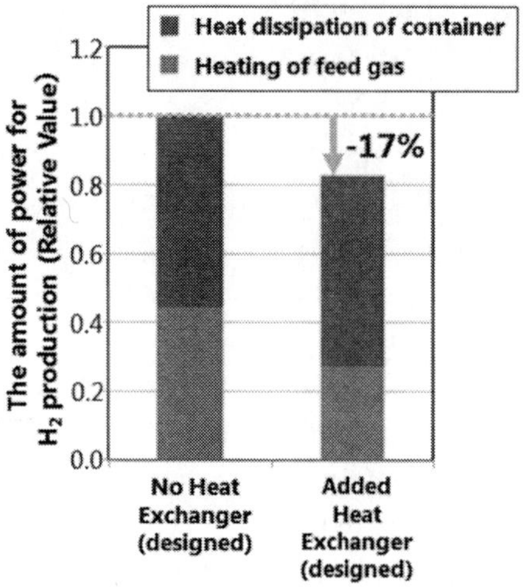

Figure 9 Comparison of power consumption during design.

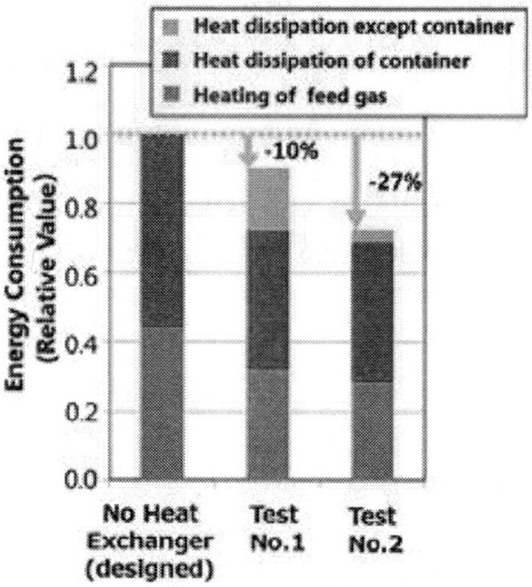

Figure 10 The test result of power consumption
Test No.1: carried out under design conditions,
Test No.2: carried out under improved operating conditions.

Figure 9 shows a comparison of power consumption under the design conditions. Here, the power consumption in the design without the heat exchanger, the process flow shown in Figure 7, was set to 1. As shown in Figure 9, by adding heat exchangers, we expected that the theoretical energy required to heat water, steam, and air will be small, and the energy can be reduced by 17%.

A comparison of the test results obtained under the design conditions and the energy consumption without the heat exchanger is shown in Figure 10. This test was carried out at steam concentration of 95%.

According to the test results, the heat dissipation from the container was smaller than the estimated value, and the energy was reduced by 10 % compared with the case without the heat exchanger.

On the other hand, from this test result, it was found that heat dissipation from the high-temperature equipment was large, for example, in the pipe between the container and the heat exchanger, and in the heat exchanger itself. Therefore, it was found that 700 °C gases cannot be supplied to the container unless it is sufficiently heated with heaters installed between the heat exchanger and the container.

From the test results obtained under the design conditions, we examined the operating method to reduce the energy consumption. Based on the test results, since the heat insulation efficiency of the container was the highest among the components of this test plant, the test was conducted to heat the gas in the container without using the heaters installed between the heat exchangers and the container. That is, the method of supplying gas at 700 °C or lower was adopted at the inlet of the container.

Although this method increases the power consumption in a container with high heat insulating performance, the temperature between the heat exchanger and the container decreases, and the amount of heat dissipation in that portion becomes small.

A comparison of the power consumption obtained with the improved operating method is shown in Figure 11. With this operating method, we could reduce the energy consumption by 27 %. From these results, we confirmed that it is possible to effectively reduce the energy consumption with a suitable equipment design and operating method.

VI. The influence of fluctuations in electrolytic power on hydrogen production

Our ultimate goal is to produce hydrogen efficiently with renewable energy. However, it is well known that renewable energy generally has an unstable output. In order to supply constant energy to the SOEC system, there is a method of storing electricity generated by renewable energy, but this method results in decreased power generation efficiency. Therefore, we conducted a hydrogen production test in which a fluctuating current was input in order to ascertain whether hydrogen production can be carried out by supplying fluctuating electricity generated by renewable energy directly to the SOEC system[4].

A simulated photovoltaic power output waveform based on an actual PV output was inputted into multi-typed cell stacks, and the response of the hydrogen production rate was evaluated. Figure 11 shows an example of actual PV output for an entire day, and Figure 12 shows the input power in this test.

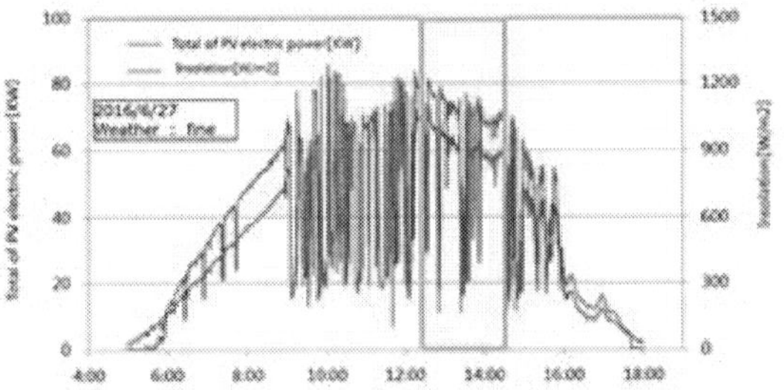

Figure 11 The example of actual PV output of all day

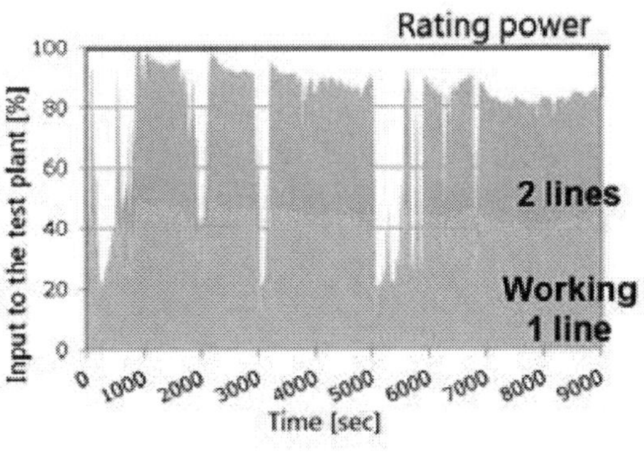

Figure 12 Input power of this test

For cell-stacks to operate effectively in the presence of fluctuations, when the input is 50% or less of the rated level, only one line works, but when the input is more than 50% of the rated, both lines work. Then, the feed gas rate is constant.

Test results of the hydrogen production rate are shown in Figure 13, and Figure 14 shows an expanded art of the test results in Figure 13. In this graph, the measured hydrogen production rate is compared with the theoretical hydrogen production rate estimated from the total current. The measured hydrogen production rate was in good agreement with the theoretical hydrogen production rate in spite of supplying a different current to each line.

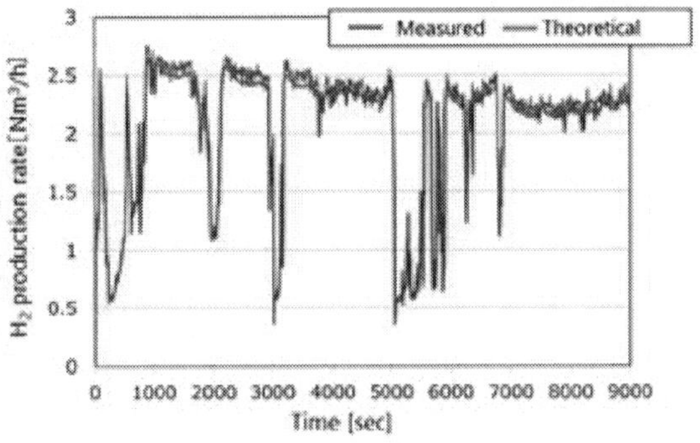

Figure 13 The test result of hydrogen production rate inputted into fluctuated current

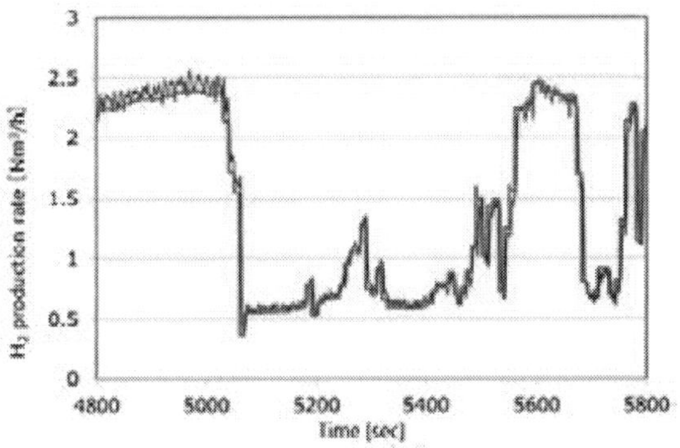

Figure 14 The test result of hydrogen production rate inputted into fluctuated current (expanded from 4800~5800 sec)

Furthermore, as shown in Figure 14, the changes in hydrogen production rate lagged the changes in current by 4–5 seconds. Since this delay corresponded to the volume from the multi-typed cell-stacks to the flow instrument, we found that the influence of the response on the hydrogen production rate of the cell-stacks by formation of multi-typed cell-stacks was very small.

VII. CONCLUSIONS

High-temperature steam electrolysis with SOEC is one of the most promising methods for hydrogen production, and has the potential to offer high efficiency. We designed a multi-typed SOEC system and carried out hydrogen production tests.

The hydrogen production tests conducted on our test plant demonstrated that by using SOEC, large-scale hydrogen production was possible even with the fluctuating energy typically produced by renewable energy.

Acknowledgement

A part of this paper is based on results obtained from a project commissioned by the New Energy and Industrial Technology Development Organization (NEDO) in fiscal 2014 to 2016. We express our appreciation.

Reference

1. M. Yoshino, K, Matsunaga, R. Nakajima, "High-Efficiency Hydrogen-Based Electric Power Storage System" *Toshiba Review Vol.70 No.5* (2015), [in Japanese]

2. K, Matsunaga, M. Yoshino, H. Watanabe, "Hydrogen Production and Hydrogen Power Storage Systems Using Solid Oxide Electrolysis Cells" *Toshiba Review Vol.71 No.5* (2016), [in Japanese]

3. N. Tsuchiya, H. Ohmura. S. Kanamura, S. Makino, M. Yoshino, S. Fujiwara, T. Kameda, M. Yamada, K. Yamada "Hydrogen Production Using Multi Solid Oxide Electrolysis Cell Stacks" *Proc. SCEJ 48th Autumn Meeting, Tokushima, Japan* (2016), Sep.6-8 [in Japanese]

4. N. Tsuchiya, H. Ohmura. S. Kanamura, M. Yoshino, S. Fujiwara, T. Kameda, M. Yamada, K. Yamada "Demonstration of Hydrogen Production by Multi-typed SOEC system (2) Influence on the system current by supply source conditions." *Proc. SCEJ 49th Autumn Meeting, Aichi, Japan* (2017), Sep.20-22 [in Japanese]

498447 A High Performing Alkaline Water Electrolyzer

Abstract:

Water electrolyzers are being developed as a way of storing renewable energy, as a way to produce hydrogen for fuel cell automobiles, and as a route to renewable fuels and chemicals. In this paper, we demonstrate alkaline water electrolyzers with iron/nickel/cobalt catalysts that show 1 A/cm^2 at 60 °C. The performance is stable for 2000 hours, showing only 3-5 μV/hr degradation. Key to the findings is the development of new Sustainion™ alkaline exchange membranes that show high conductivity and good stability in 1 M KOH.

These results show that the use of Sustainion™ membranes allow one to greatly improve the performance of alkaline water electrolyzers. While the initial voltage is higher than a typical PEM electrolyzer with platinum/iridium catalysts, the stability is better, so after 1000 hours or so, the performance is quite similar. This presents the possibility of one replacing rare materials such as iridium with iron and nickel, and still maintaining high currents and the ability to turn the system on and off quickly.

501248 Iridium Nanowires As Highly Active, Oxygen Evolution Reaction Electrocatalysts

Abstract:

Iridium Nanowires as Highly Active, Oxygen Evolution Reaction Electrocatalysts

Shaun M. Alia, Sarah Shulda, Chilan Ngo, Svitlana Pylypenko, Bryan S. Pivovar

In the United States, 2% of energy used goes through the hydrogen pathway, to produce ammonia in agriculture and upgrade oil in transportation. [1, 2] While most hydrogen is produced by steam methane reformation, electrolysis can become competitive with the use of low-cost renewable power sources. Further reductions in the cost of renewables may increase hydrogen use as an energy intermediate between the electric grid and transportation and industrial processes. [3]

Electrolyzers today operate at high capacity and with constant power input. A shift toward intermittent renewables and a focus on hydrogen production cost, however, increases the importance of catalyst thrifting and durability in the oxygen evolution reaction (OER). [2] Iridium or iridium oxide nanoparticles are typically used as OER catalysts, but are limited in activity and durability, particularly at low loading. [4]

Iridium-nickel and iridium-cobalt nanowires have been developed as OER electrocatalysts for proton exchange membrane-based electrolyzers. These catalysts use similar templates and synthesis routes previously used in the development of fuel cell oxygen reduction electrocatalysts. [5] A previous comparison between polycrystalline iridium and nanoparticles suggests that extended structures can potentially benefit from higher site-specific activity. [4] Differences between the nanowire templates affect catalyst composition, structure, and OER activity. While these materials exceed the performance of iridium nanoparticles, acid leaching is necessary to minimize template dissolution and improve durability. The acid leached catalysts exceed the half-cell activity of nanoparticles by an order of magnitude, and the half-cell mass activity of any catalyst available in literature by 4 times. In membrane electrode assemblies, the nanowires also outperform traditional iridium catalysts by 4â€'5 times in single-cell electrolyzers.

[1]Â A. Milbrandt and M. Mann, ed. U. S. Department of Energy, http://www.nrel.gov/docs/fy09osti/42773.pdf, 2009.

[2] U. S. Department of Energy, https://www.hydrogen.energy.gov/pdfs/review16/2016_amr_h2_at_scale.pdf, 2016.

[3] P. Denholm, M. O'Connell, G. Brinkman and J. Jorgenson, ed. U. S. Department of Energy, http://www.nrel.gov/docs/fy16osti/65023.pdf, 2015, vol. NREL/TP-6A20-65023, ch. NREL/TP-6A20-65023.

[4] S. M. Alia, B. Rasimick, C. Ngo, K. C. Neyerlin, S. S. Kocha, S. Pylypenko, H. Xu and B. S. Pivovar, Journal of The Electrochemical Society, 2016, 163, F3105-F3112.
[5] S. M. Alia, Y. S. Yan and B. S. Pivovar, Catalysis Science & Technology, 2014, 4, 3589-3600.

499657 The Effect of Two Different Preparation Method of Cu@Pd/C in Direct Formic Acid Fuel Cell

Abstract:

DFAFCs use anode catalysts based on precious metals such Pd. Unfortunately, palladium has poor stability, and it can be poisoned easily by carboxy species like $C(OH)_2$. The catalyst poisoning leads to poor stability and efficiency of DFAFC. Therefore, it is important to develop catalysts that are used in the anode of DFAFC. By using the core-shell structure of non-expensive 3d transition metals as core and the Pd as the shell, an electronic structure like the d-band center can be modified by charge transfer effects; additionally, the selectivity and activity will be altered. This lead to changes in adsorption energy of intermediate species during the electro-oxidation of formic acid on Pd surfaces. Because of using these 3d transition metals as a core and Pd as a shell in our core-shell structures, the bond energy between Pd and the intermediate formed in the formic acid electro-oxidation can be formulated with optimum bonds, which is strong enough to cause adsorption not too strong to produce poisoning species.

The studies on the core-shell bimetallic catalysts have resulted in better fuel cell performances, and different electronic properties correspond to pure Pd. Among the Core-shell structure bimetallic catalysts Cu@Pd showed the better performance. However, there are lots of preparation methods to produce the Cu@Pd core-shell catalysts, and no one has compared them yet. In the present study, Cu@Pd/C were prepared by two different methods. Core-shell and skeleton structures of Cu@Pd/C nanoparticle samples were prepared with the average size of the 6.34± 0.74 nm and 6.17± 0.31 nm, respectively. XPS data illustrate that the Pd $3d_{5/2}$ shifts positively to higher binding energy, and the d-band center moves away from the Fermi level. Changing in the BE and d-band center shows that the electronic structure of the surfaces were modified which resulted in higher activity and stability than the Pd/C toward the FAO.

500930 Enhanced Performance with Ni-Based Single Solid Oxide Fuel Cells with Mo-CZ Composition

Abstract:

This work is focused on investigating an alternative anode material that can effectively mitigate carbon deposition when complex hydrocarbon fuel (e.g. isooctane) is directly fed to a solid oxide fuel cell (SOFC). Previous studies have demonstrated that a Ni-based cermet layer can provide good heterogeneous catalytic activity and electrochemical activity towards oxidation of hydrocarbon fuel. However, severe carbon deposition from cracking of hydrocarbons on the nickel phase leads to cell degradation and poor performance stability. Thus, the goal of this research is to maintain high catalytic performance of Ni, while suppressing carbon deposition by introducing Mo to the cermet layer.

In this research, the key processing technology for fabricating the bottom single cell with NiMo-Ceria Zirconia (NiMo-CZ) anode has been developed. Electrochemical performance and performance stability of the button cell were investigated with isooctane at 800 °C. As compared with a conventional Ni/YSZ anode single cell, the Ni-CZ anode single cells showed distinctly improved performance stability. This result indicates the advantage of using CZ in the anode, which has high oxygen storage and release capacity (OSC), by efficiently oxidizing carbon formed on active sites and mitigating carbon deposition. By increasing Mo content from 0% to 3% and 5%, the maximum power density significantly increased from 80 $mW \cdot cm^{-2}$ to 120 $mW \cdot cm^{-2}$ and 240 $mW \cdot cm^{-2}$, respectively, under the same operating conditions. This suggests that increased heterogeneous catalytic activity toward isooctane partial oxidation with the addition of Mo.

Furthermore, NiMo-CZ (5% Mo) exhibited excellent performance stability at a constant current density of 150 $mA \cdot cm^{-2}$ under the direct feed condition of isooctane. Within the first 3 hours, cell voltage degradation was observed due to the change of fuel from H_2 to isooctane. In the following next 48 hours, the cell voltage degraded at a rate of 3.4 $mV \cdot h^{-1}$, while degradation rates of 4.6 $mV \cdot h^{-1}$ were observed for NiMo-CZ (3% Mo) and 5.7 $mV \cdot h^{-1}$ were observed for Ni-CZ. The results from these measurements suggest that the incorporation of additional Mo further suppresses carbon deposition on anode. When Mo is introduced, it could alter the electric property of Ni surface, which decreases the bond strength between the Ni and C. The results demonstrate the feasibility of using NiMo-CZ (5% Mo) as a promising SOFC anode when complex hydrocarbons are directly fed to the system.

Teaching the Need to Account for Mixing in th Design of Tubular Reactors

Recommended by Stuart W. Churchill

Warren D. Seider
Department of Chemical and Biomolecular Engineering
University of Pennsylvania

Oct. 30, 2017

In the Chemical Flowsheet Simulators, the common tubular reactor model is –

i.e., perfect radial mixing with no wall drag

For many years Prof. Stuart Churchill questioned the adequacy of the model –

Yields high reactor conversions

Predicts shorter reactor lengths for specified conversions

Also, apparently all chemical reactor design textbooks present the PFR assumption without discussing its inadequacy for many design calculations; e.g.,

Fogler
Levenspiel
Rawlings and Eckert
Hill

Some account for back-mixing by adding partial recycle about PFR models; i.e.,

In response:

Seider et al., *Product and Process Design Principles*, 4th Ed., Wiley, 2017

Chapter 15, Chemical Reactor Design

Begin with saponification reaction:

$$A + B \rightarrow C + D$$
$$EtAc + NaOH \rightarrow NaAc + EtOH$$

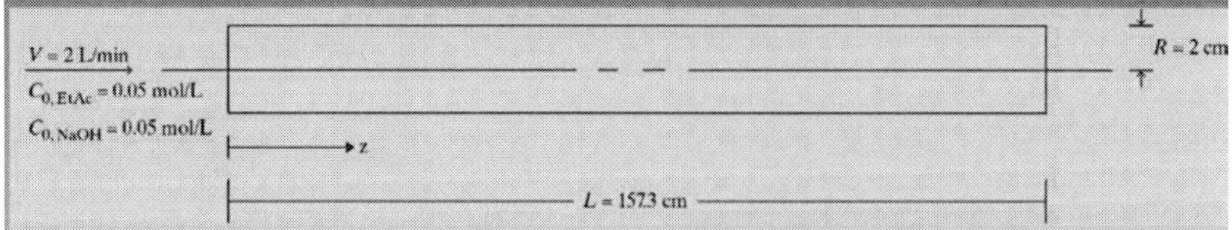

$$X_{\text{PFR}} = \frac{1}{\left(1 + \frac{u_{\text{mean}}}{k c_{\text{A0}} L}\right)} = 0.306$$

But, for Laminar Flow:

$$Re = 1{,}323$$

$$\frac{u\{R\}}{u_{\text{mean}}} = 2\left(1 - \left(\frac{R}{R_t}\right)^2\right)$$

$$X_{\text{LD0}}\{R\} = \frac{1}{\left(1 + \dfrac{2u_{\text{mean}}\left(1 - \left(\dfrac{R}{R_t}\right)^2\right)}{kc_{\text{A0}}L}\right)}$$

$$\overline{X}_{\text{LD0}} = 2\pi \int_0^{R_t} X_{\text{LD0}}\{R\}\, u\{R\}\, R dR$$

$\overline{X}_{\text{LD0}} =$ 0.274 ~10% lower

Teaching Mixing in the Design of Tubular Reactors 5

Helical Flow:

Feed at 0.05 mol/L of sodium hydroxide and 0.05 mol/L of ethyl acetate.

Minor radius at 2 cm (0.7874 in)

Major radius at 25 cm (9.8425 in)

Pitch at 5 cm (1.9685 in)

Temperature is 30°C.

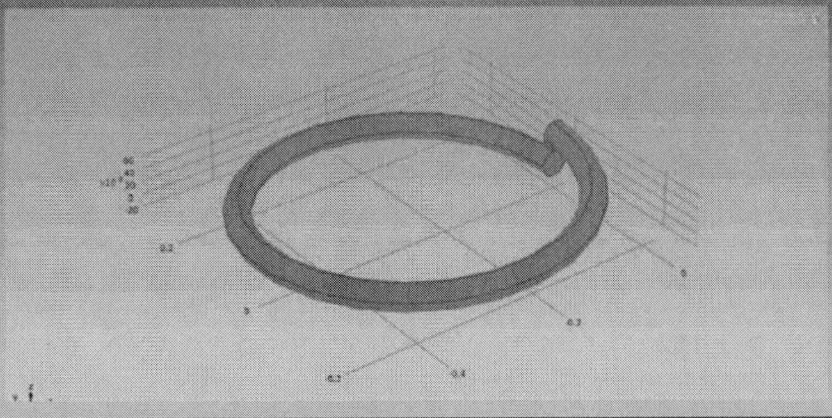

Teaching Mixing in the Design of Tubular Reactors 6

Reactor Exit

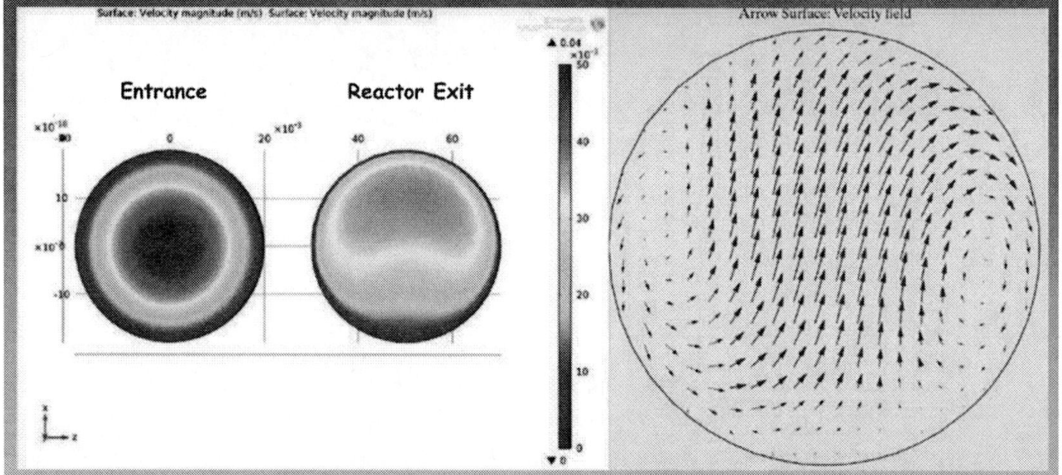

$\bar{X}_{LD0} =$ 0.304 < 0.306 (PFR)

Teaching Mixing in the Design of Tubular Reactors 7

Non-Isothermal Straight Tubular Reactor - Fully-Developed Laminar Flow

Adiabatic

Arrheneus Equation

$$k = Ae^{-\frac{E_a}{RT}}$$

Species	ΔH_f^o(kJ/mol)	C_p(kJ/mol-K)
Ethyl acetate	−480.3125	0.169
Sodium hydroxide	−469.15	0.08785
Ethanol	−276	0.112
Sodium acetate	−709.32	0.10083

Table 15.5 Equations for Heat Transfer and the Boundary Conditions

Condition	Equation	Applicable Area
Heat transfer	$\rho^* Cp^* u \cdot \nabla T = \nabla \cdot (k\nabla T) + Q + Qvh + Wp$	Domain 1
Temperature	$T = T0$	Boundary 2
Outflow	$-n \cdot (-k\nabla T) = 0$	Boundary 3
Insulation	$-n \cdot (-k\nabla T) = 0$	Boundary 4
Heat source	$Q = [A^* \exp(-Ea/R/T)]^* c\hat{\ }2^* dH$	Domain 1

Teaching Mixing in the Design of Tubular Reactors 8

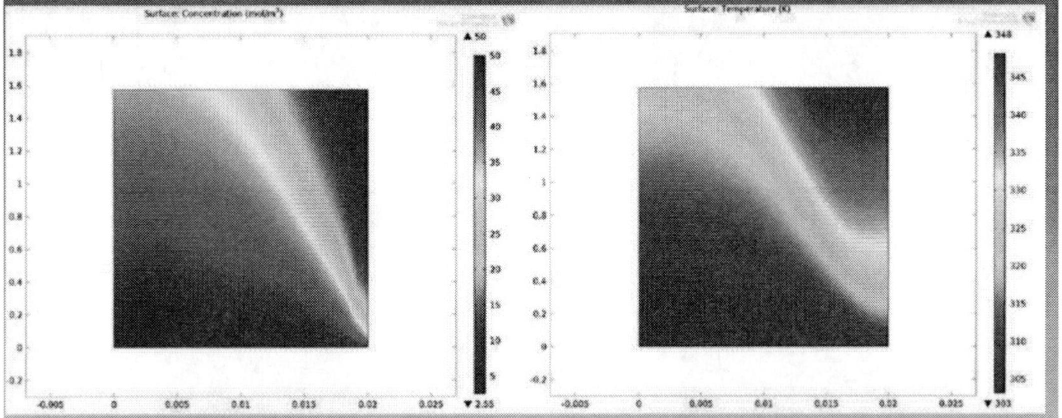

$\overline{X}_{LD0} =$ 0.52 (much higher than 0.274 at 30°C)

Teaching Mixing in the Design of Tubular Reactors 9

Non-Isothermal - Diabatic

Fully-Developed Laminar Flow with Radial Diffusion

Uniform wall T = 303.15 K

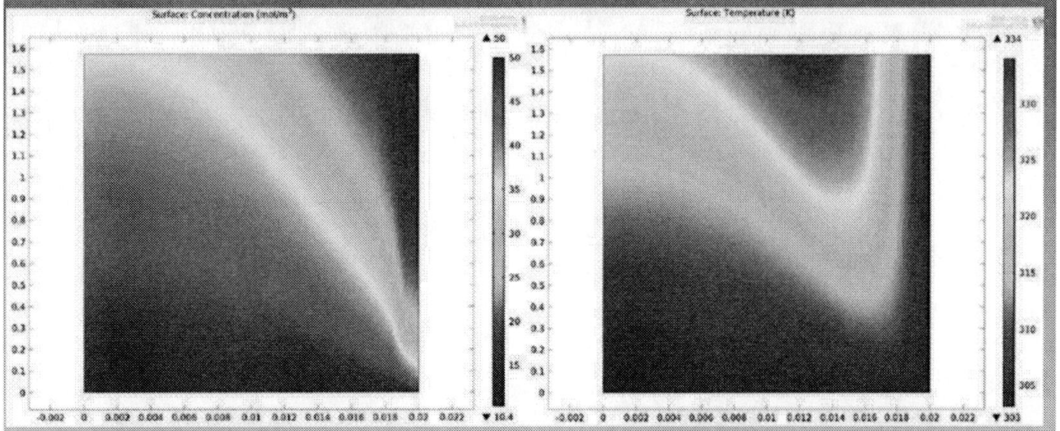

$\overline{X}_{LD0} = 0.461$

Teaching Mixing in the Design of Tubular Reactors 10

Non-Isothermal - Diabatic

Plug Flow

Uniform wall T = 303.15 K

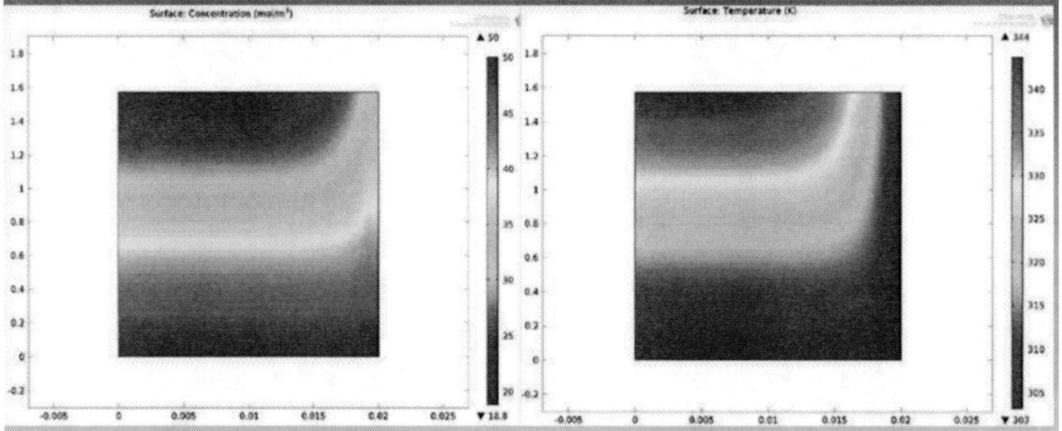

$$\overline{X}_{\text{LD0}} = 0.544$$

Teaching Mixing in the Design of Tubular Reactors 11

Using COMSOL, Chapter 15 shows how to:

- specify the reactor geometry
- enter the differential equation model
- enter the boundary conditions
- create the finite-element mesh structure
- interpret and graph the results – often with considerable post-processing

Chapter 15 also presents several homework exercises, including one on confined-jet mixing –

the subject of my doctoral thesis
suggested by Prof. Churchill in Sept. 1962!

Teaching Mixing in the Design of Tubular Reactors 12

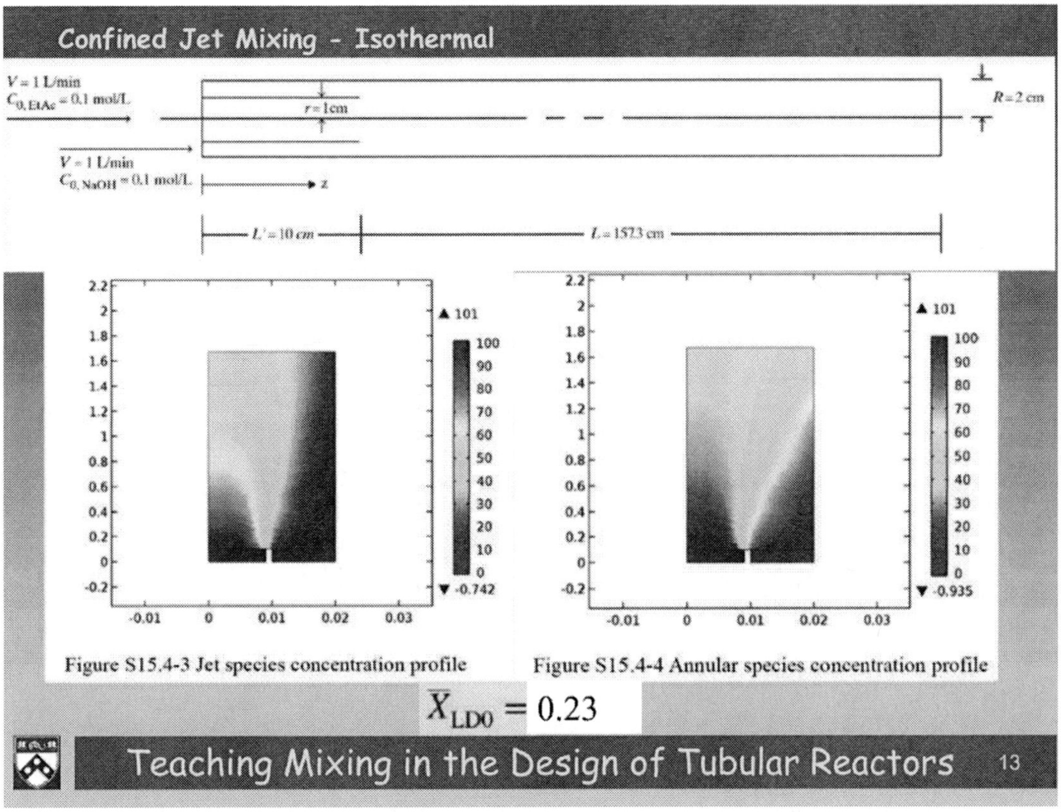
Confined Jet Mixing - Isothermal
V = 1 L/min
C0,EtAc = 0.1 mol/L
V = 1 L/min
C0,NaOH = 0.1 mol/L
r = 1cm
R = 2 cm
z
L' = 10 cm
L = 1573 cm
▲ 101
▼ -0.742
▲ 101
▼ -0.935
Figure S15.4-3 Jet species concentration profile
Figure S15.4-4 Annular species concentration profile
X̄LD0 = 0.23
Teaching Mixing in the Design of Tubular Reactors
13

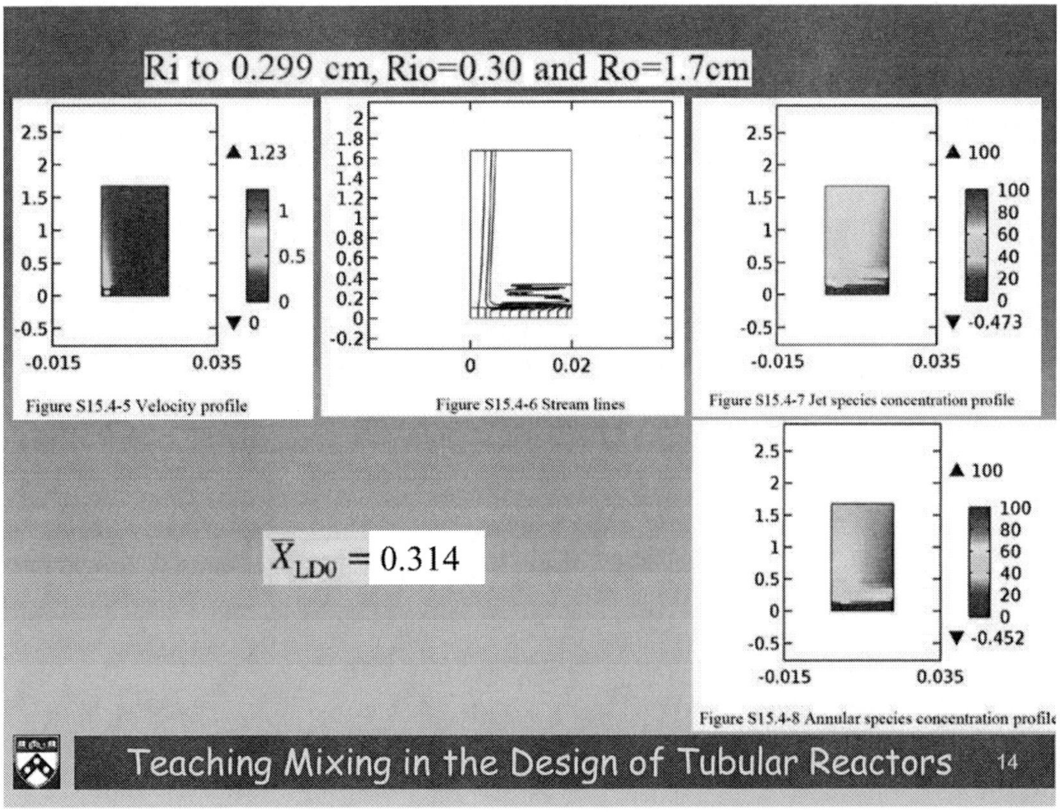
Ri to 0.299 cm, Rio=0.30 and Ro=1.7cm
▲ 1.23
▼ 0
Figure S15.4-5 Velocity profile
Figure S15.4-6 Stream lines
▲ 100
▼ -0.473
Figure S15.4-7 Jet species concentration profile
X̄LD0 = 0.314
▲ 100
▼ -0.452
Figure S15.4-8 Annular species concentration profile
Teaching Mixing in the Design of Tubular Reactors
14

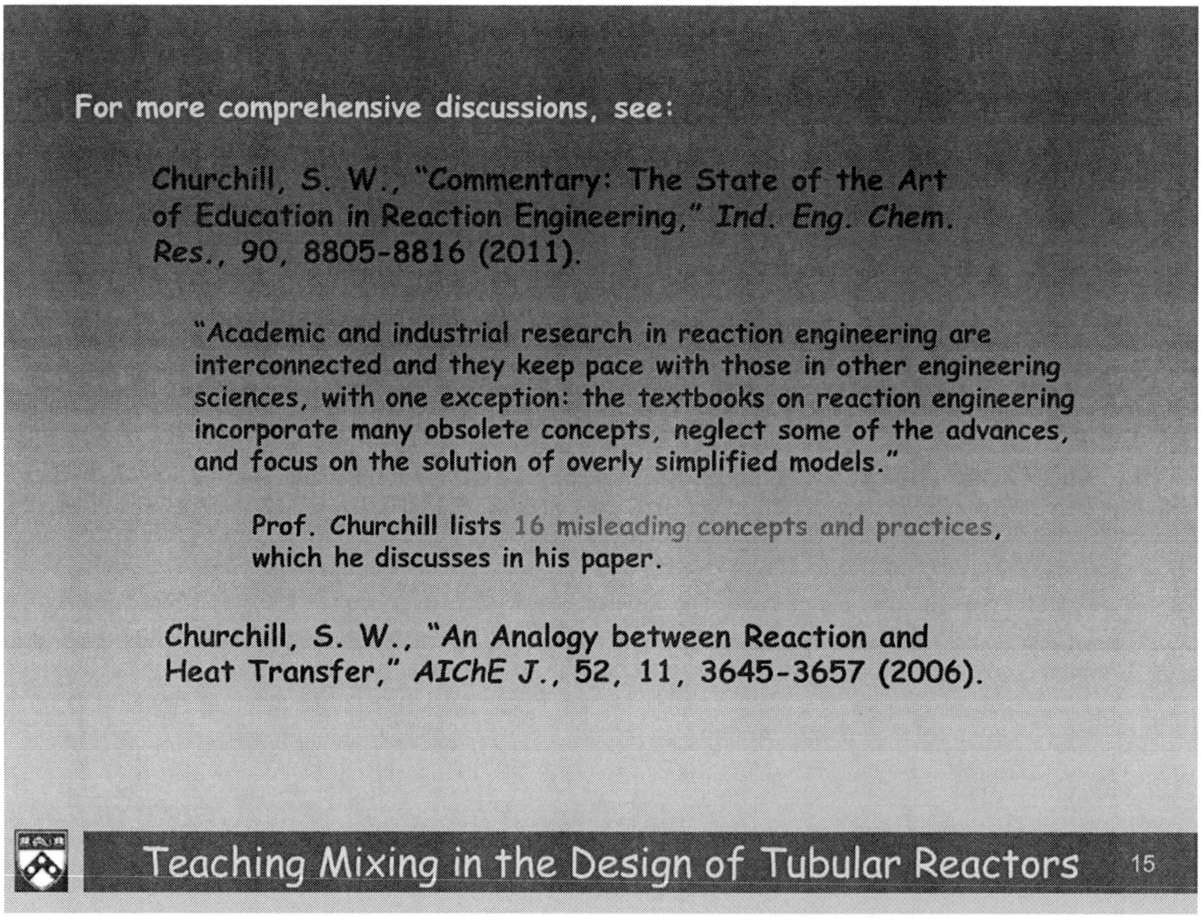
For more comprehensive discussions, see:
Churchill, S. W., "Commentary: The State of the Art of Education in Reaction Engineering," *Ind. Eng. Chem. Res.*, 90, 8805-8816 (2011).
"Academic and industrial research in reaction engineering are interconnected and they keep pace with those in other engineering sciences, with one exception: the textbooks on reaction engineering incorporate many obsolete concepts, neglect some of the advances, and focus on the solution of overly simplified models."
Prof. Churchill lists 16 misleading concepts and practices, which he discusses in his paper.
Churchill, S. W., "An Analogy between Reaction and Heat Transfer," *AIChE J.*, 52, 11, 3645-3657 (2006).
Teaching Mixing in the Design of Tubular Reactors
15

503676 Clustering of Heavy Particles in Turbulence - Dimensional Analysis and Scaling Laws in the Inertial Range

Abstract:

High-resolution direct numerical simulations of small particles embedded in a turbulent flow field show that the particles accumulate in regions of high strain and low vorticity. While clustering is more dramatic at small scales, it is also observed throughout the inertial range of turbulence. We present a statistical theory for the scaling law of the two-particle distribution function in the inertial range. The scaling is developed with the aid of dimensional analysis similar to that taught by Professor Churchill in his graduate course on heat transfer, which laid out the simple, yet profound guidance obtained from the constraints imposed by the units of the relevant variables. Even in this completely different application, those insights I learned 30 years ago played a vital role in helping us understand and model clustering in the inertial range.

503677 Measurements and Prediction of Sooting Tendencies of Hydrocarbons and Oxygenated Hydrocarbons

Abstract:

One of Stuart Churchillâ€™s lessons is that experimental measurements should be cast in terms that can be generalized to provide information to other systems and conditions than those of the initial measurements. The amount of soot formed in combustion devices such as engines depends strongly on the chemical structure of the fuel. Over the past 10 years, we have measured quantitative sooting tendencies for hundreds of pure hydrocarbons representative of those in practical fuels. The sooting tendencies have been characterized in terms of Yield Sooting Index (YSI), which is based on the amount of soot formed in a methane/air reference flame when a small amount of the test compound is doped into the fuel. In this new collaborative project, we are developing a Quantitative Structure-Activity Relationship (QSAR) model from these results; this model can be used to predict YSI for new hydrocarbons of interest and to understand the details of the underlying chemistry. The model uses molecular descriptors to describe the structure of each compound in the database, and it predicts YSI with a feed-forward artificial neural network trained with the measured values. It has a rigorously defined applicability domain and its performance has been tested by both internal and external validation. New measurements have been performed to extend the applicability domain towards advanced biofuels and for external validation. To demonstrate that YSI is relevant to practical systems, we have also been measuring YSI for mixtures of promising biofuels and conventional gasoline, and comparing these to particulate emissions measurements from gasoline direct injection (GDI) engines [Ratcliff et al., *SAE Int J Fuels Lubr doi* :10.4271/2016-01-0705]. The results show that YSI and engine emissions are related, but that other factors such as fuel volatility also play a role.

503678 Role of Brain Interstitial Transport in Alzheimer's Disease

Abstract:

Alzheimerâ€™s disease (AD) is the principal cause of dementia in the US, affecting 5 million individuals, and is one of the fastest growing diseases in other developed countries, afflicting 38 million worldwide. The recent findings of the presence of lymphatic vessels in the subarachnoid space of the brain strongly support exchange between the cerebrospinal fluid (CSF) and the interstitial fluid that envelopes neurons and glia in the brain. This opens new opportunities to study the diffusive and convective transport of harmful products from the brain. The analysis of transport phenomena, as so profoundly developed and taught by Professor Churchill, applies broadly, even in these non-traditional problems.

AIChE ANNUAL MEETING MINNEAPOLIS

Biographical Perspectives of Stuart W. Churchill

Warren D. Seider
University of Pennsylvania
Department of Chemical and Biomolecular Engineering
Philadelphia, PA 19104

October 30, 2017

Sept. 1938 – Stuart began studying undergrad mathematics at UM –
received much advise from his freshman professor, Clyde Love

During next 4 years – took 'all' possible ChE courses –
worked on retrograde phase equilibria in Prof. Don Katz's lab – published a paper

As a graduating senior, he applied for and received a BSChE Degree

May 1942 – accepted a position at Shell Development Corp. – Wood River, IL Refinery – worked on new processes such as catalytic cracking, and aviation gasoline-related problems to assist in WWII

Sept 1947 – returned to UM for doctoral studies in ChE

1950 – while doing doctoral research, was asked by Profs. G.G. Brown and D. L. Katz to be a ChE Instructor

1950 - Graduate Student - Teaching at UM

Remarkable 10-Year Acceleration

1952	Defended his doctoral thesis
1956-1960	16 of his doctoral students defended their theses!!!

Turbulent Flow
Light Scattering
Dimensional Analysis
Natural Convection and Combustion*
- all with limitations due to slow, small computers

1957	Promoted to Full Professor

* To appreciate these and later contributions, read Stuart's paper in the *J. Heat Transfer*, ASME, "A Parallel Universe: Contributions to the Initial Development of Computational Heat Transfer", which refers to the parallel work of F.H. Harlow and D.B. Spalding"

In the next five years:

1962	Chair Dept. of Chem. and Met. Eng. (~30 faculty)
1963	Vice-Chair Board of Control of Intercollegiate Athletics
1963	Vice-Chair of Senate Advisory Committee on University Affairs
1964-1966	V. President, President, P. President of AIChE

I encountered Stuart for the first time in Sept. 1962 at his 8:00 A.M., MWF, Rate Processes Course.

Typically, by 8:05 A.M., he had one entire blackboard filled with a mathematical model - and was beginning to fill a second board. He wrote quickly without notes.

Stuart's approach to teaching transport phenomena actively involved his students with instantaneous derivations at the blackboard in response to their questions - often erasing with his left hand while writing with his right hand.

Three highlights

As Chair - Dept. of Chem. and Metallurgical Eng.
with President Harlan Hatcher -
- receiving a gift from Dale Biggs (a former classmate), Dow Chem.

As 1965 AIChE President

with ICE President, P. V. Danckwerts

Countered the Threat

ASEE preliminary report on
Goals of Engineering Education
Called for –

- Five year professional degree for all engineers – finish with MS
- Change from specific curricula accreditation to accreditation of the overall engineering college

Implied – "elimination of chemistry and its applications"
i.e., chemical engineering

In 1967, Stuart accepted the Carl V. S. Patterson Professorship at UPenn –

to refocus his emphasis on teaching and research

He wrote the textbook that he and R.R. White initiated with their lectures at UM in the late 1950s.

Introducing the rate concept; i.e., the distinction between 'rates of change' and 'process rates', focusing on rate data

Famous Quotes (~200) Interspersed Throughout

The Rate Concept (p8)

"The physical, chemical, and biological transformations of concern to engineers must be completed in a reasonable period of time. The time required for specific change is inversely proportional to the rate at which the process occurs."

Quote: "Remember that time is money" - Benjamin Franklin

Potential Factors and Rate Coefficients (p175)

"Experience with rate processes of all kinds has indicated that one convenient way to formulate relationships between the rate and the environment is to separate 'potential factors' which drive the rate processes from 'resistance factors' which oppose the processes."

Quote: "Inequality is the cause of all local movements" - Leonardo da Vinci

SWC - a Renaissance man - integrating arts, politics, science

With faculty and students, he exercised

We won the first annual F.J. Van Antwerpen AIChE 4-mile Relay Championship!

(UPenn defeated UColorado - K. Timmerhaus, M. Peters, ...)

UPenn Annual Canoe Trip - Delaware River
Honored Guests - R. Byron Bird, E. Lightfoot

He participated in nearly all AIChE meetings
- 1978, in Chicago

Awards came in rapid succession

1964	AIChE Professional Progress Award
1969	AIChE William H. Walker Award
1974	Elected to National Academy of Engineering
1978	AIChE Warren K. Lewis Award
1979	ASME-AIChE Max Jacob Award in Heat Transfer
1980	AIChE Founders Award
.	
.	
.	
2002	NAE Founders Award (1/yr - previous 50 years)

1980 Philadelphia Summer Meeting of AIChE

60th Birthday Celebration
Upper Egyption Room – UPenn Museum

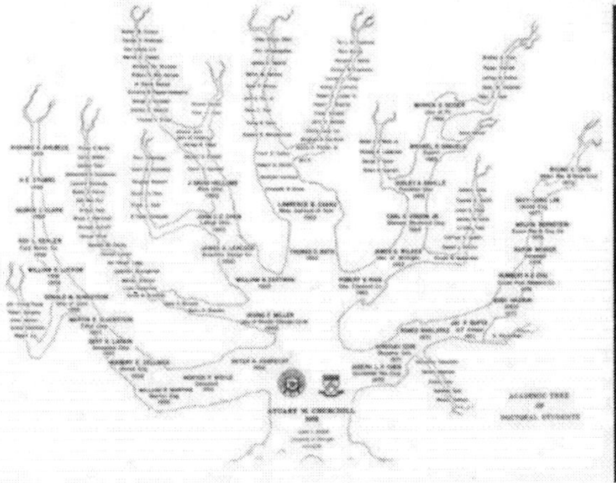

Oral History - Chemical Heritage Foundation - June 1985

Interview by Joseph C. Marchese and Jeffrey L. Sturchio

Excellent accounting of first 45 years of his professional career

Key themes

Mathematical training positioned him to play leading role in modeling transport phenomena, chemical kinetics, combustion, ...

Industrial experience prior to doctoral research gave him an appreciation of process operations, process design (i.e., CAD - he understood the challenges of my research), ...

Early research - funded mostly (90%) by Defense Dept.
e.g., role of scattering of thermal radiation from nuclear explosions

Overcoming challenges of modeling with slow, small computers

80th Birthday - Portrait Unveiled, Libby Rudnick, Artist

In London, Stuart saw the play Copenhagen by Michael Frayn – portraying the meeting in 1941 in Copenhagen among Werner Heisenberg, and Niels and Margrethe Bohr.

In 1939, as a sophomore, Stuart had observed the conference of atomic scientists at UM involving Samuel Goudsmit, Clarence Yoakum (UM VP), Werner Heisenberg, Enrico Fermi, and Edward Kraus (UM Dean) – also, Niels Bohr

In Sept. 1999, when writing to entering Penn freshmen, all of whom read the book Copenhagen before classes began, Stuart wrote:

"I found the play in London engaging and thought-provoking. I hope you had a similar reaction in reading it this summer.

Although the play focuses on a particular occasion, it is really about conflicting loyalties to one's family, one's friends, one's professional peers, one's professional reputation, one's country, and humanity.

I warn you that the play is not a reliable guide to atomic physics or to the history of atomic reactors and weapons. As I listened to the play, alarm bells kept ringing in my mind because of incongruities in history and science ..."

"... I will tell you why I have read so much on this subject, because it is very relevant to your careers at Penn and perhaps later in graduate and professional school."

Old Wine Dinners – **with many of his 45 former doctoral students at all Annual AIChE Meetings**

When he died at age 95, he was writing another manuscript for *Ind. Eng. Chem. Res.* (his 110th paper since forced retirement at age 70 in 1990):

"Improved Algebraic, Numerical, and Graphical Representations in Fluid Mechanics"

A sample page follows – Stuart typed the manuscript, as well as the equations

4. TURBULENT FLOW IN A ROUND TUBE

Osborne Reynolds[8] in 1895 made the greatest single advance of all time in the description of turbulent flow in a round tube when he time-averaged the partial differential equations that describe the conservation of mass, momentum, and energy and thereby reduced them to a manageable form. He actually space-averaged the equations but despite quibbling by purists, no serious consequential error has ever been identified that resulted from that transformation. Before time-averaging he represented each variable by the sum of a steady state and a fluctuating component and worked out all the algebra that ensues. The result for fully developed flow is

$$\tau = \mu\frac{du}{dy} - \rho\overline{u'v'} \qquad (18)$$

where

$$\tau = \tau_w\left(1 - \frac{y}{a}\right) \qquad (19)$$

In that $\overline{u'v'}$ was found by Reynolds by means of both a purely theoretical analysis and all subsequent conductors of experimental measurements to be a negative quantity, the shear stress within the fluid is revealed to be the sum of the contributions of the laminar shear stress and that due to turbulence. As re-expressed in wall-based notation, Eq. 18 becomes

$$1 - \frac{y^+}{a^+} = \frac{du^+}{dy^+} + \left(\overline{u'v'}\right)^+ \qquad (20)$$

Pete Lederman finishing his introduction of Stuart at a reception - SF Annual Meeting of AIChE, Nov. 2013

502794 Vapor Recompression Mechanism Introducing in Internal Heat-Integrated Distillation Column: Impact of Internal Energy Driven Intermediate and Bottom Reboiler

Abstract:

A novel combination of internally heat-integrated distillation column (HIDiC) and vapor recompression column (VRC) with intermediate reboiler (IR) is proposed. Supplying heat at the highest temperature point (i.e., column bottom) of the VRC scheme is not thermodynamically favorable and, therefore, we aim to install the IR for better distribution of heat along the column length, thereby reducing the compressor work. Introducing IR in the combined HIDiC-VRC system formulates an open-loop variable manipulation policy to evaluate the comparative impact of internal and external heat sources on bottom liquid reboiling. With internal energy driven bottom reboiler, we further investigate the hybrid HIDiCVRCIR column with proposing the two modes of compressor arrangement, namely parallel and series. Finally, a multicomponent distillation system is exampled to show the promising potential of the proposed HIDiC-VRCIR configurations in improving the energetic and economic performance over the HIDiC-alone and HIDiC-VRC schemes with reference to a conventional standalone column.

496355 Experimental Studies Controlling Trace Components in a Dividing Wall Distillation Column

Abstract:

Experimental Studies Controlling Trace Components in a Dividing Wall Distillation Column

Melissa Donahue, Michael Baldea, R. Bruce Eldridge

McKetta Department of Chemical Engineering and

James R. Fair Process Science and Technology Center

The University of Texas at Austin

An example of one of the prominent technologies in the field of process intensification is dividing wall distillation columns or DWCs. As part of an ongoing experimental study of dividing wall columns at The University of Texas at Austin[1], the design and selection of control strategies for a pilot plant DWC will be presented. The pilot plant column is six inches in diameter and approximately 30 feet tall. The column has 24 theoretical stages of structured packing and a welded wall in the horizontal and vertical middle of the column[1]. To investigate the operational flexibility of this fixed design DWC, a fourth trace component is added to the feed. A sensitivity analysis studying the minimum energy usage and optimal distribution of impurities is conducted using a dynamic process simulator. These operating targets and conventional distillation temperature control techniques are used to design control configurations that are then implemented on the pilot scale dividing wall column. The control configurations are tested against changes in set point and feed flow and composition disturbances.

(1) Blevins, T.; Downs, J. J.; Donahue, M. M.; Roach, B. J. Use Model Predictive Control to Achieve Real-Time Management of a DWC. *Hydrocarbon Processing* **2015**.

UNIVERSITY OF BATH

CASCADE REACTION IN A SPINNING MESH DISC REACTOR USING A NOVEL METAL- ENZYME CATALYST

PARIMALA SHIVAPRASAD
Department of Chemical Engineering
ps784@bath.ac.uk

Emma Emanuelsson*, Darrell Patterson*, Matthew Jones**
* Department of Chemical Engineering, University of Bath
** Department of Chemistry, University of Bath

Bath - A University in a UNESCO World Heritage City

UNIVERSITY OF BATH

Outline

UNIVERSITY OF BATH

Background

The Idea

Results

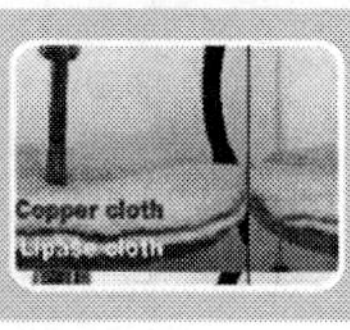

The Path Forward

Problem

UNIVERSITY OF BATH

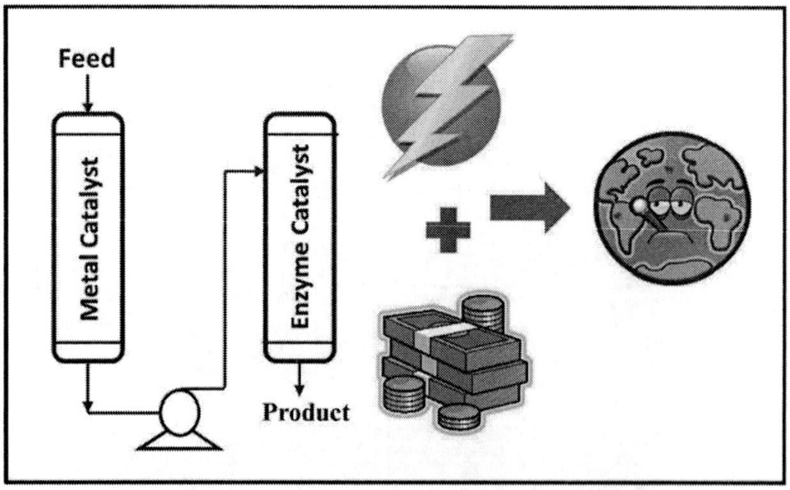

Multi-step processes lead to higher recovery and energy costs.

Solution

UNIVERSITY OF BATH

Imagine a reactor that could do a cascade of reactions without the need for product purification in between!

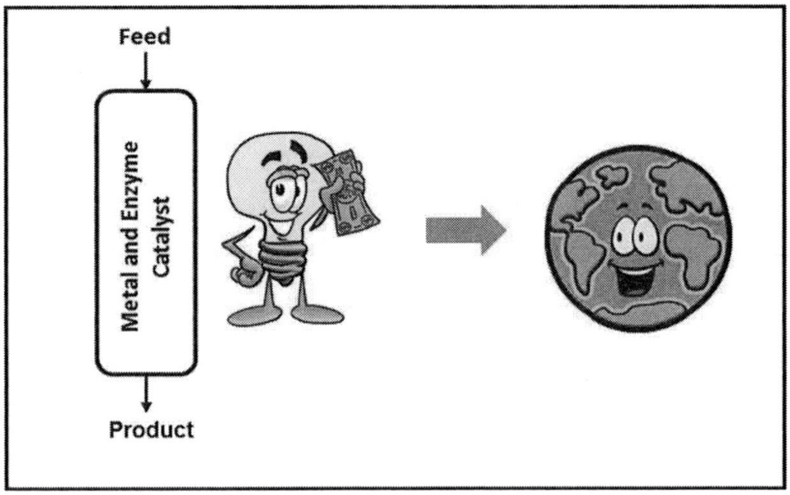

Catalyst immobilisation on wool

UNIVERSITY OF BATH

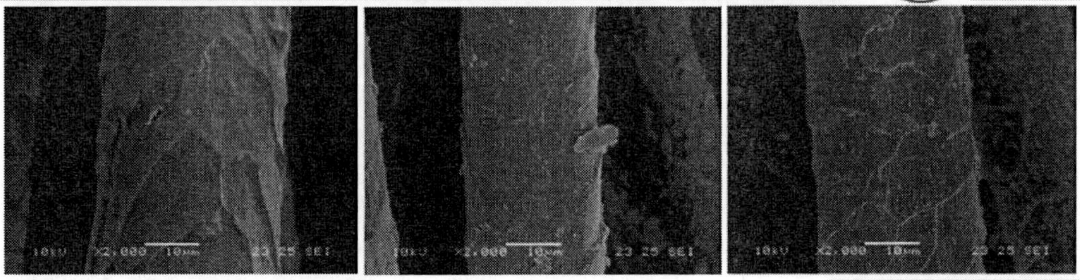

Shivaprasad, P., *et al.(2017) Chemical Engineering and Processing: Process Intensification.*

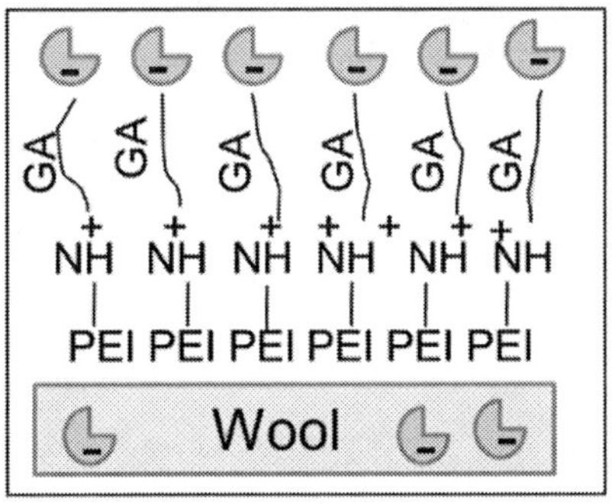

Feng, X., *et al.(2012) Colloids and Surfaces B: Biointerfaces*, 102, pp. 526-533

Bleached wool

↓ **2% PEI, 2h**

Wool-PEI

↓ **Catalyst (Metal or enzyme), 24 h**

Wool-PEI-catalyst

↓ **0.5% GA,10 min**

Wool-PEI-catalyst-GA

The Idea: Process Intensification

UNIVERSITY OF BATH

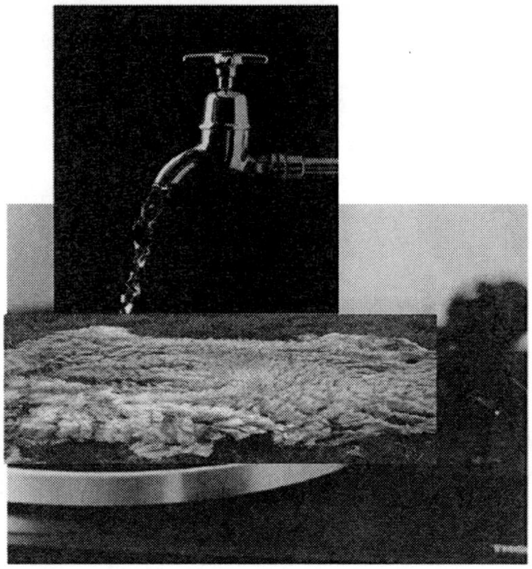

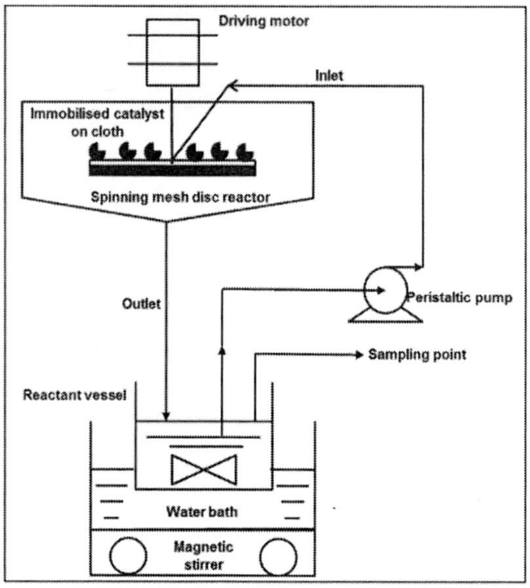

Spinning Mesh Disc Reactor (SMDR)

The SMDR: Characteristics

UNIVERSITY OF BATH

Reactor characterisation for enzymatic hydrolysis of tributyrin

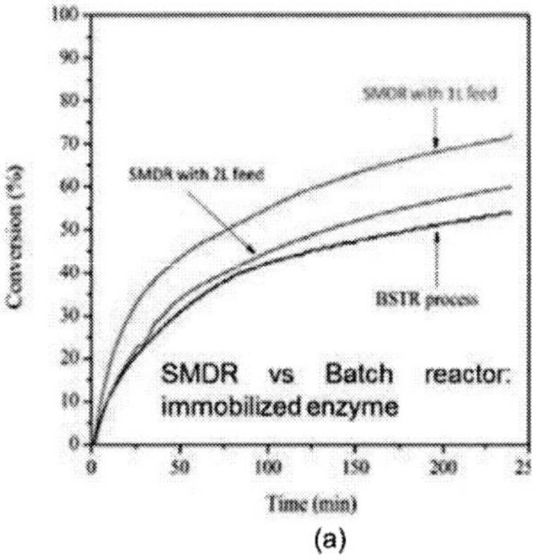

(a)

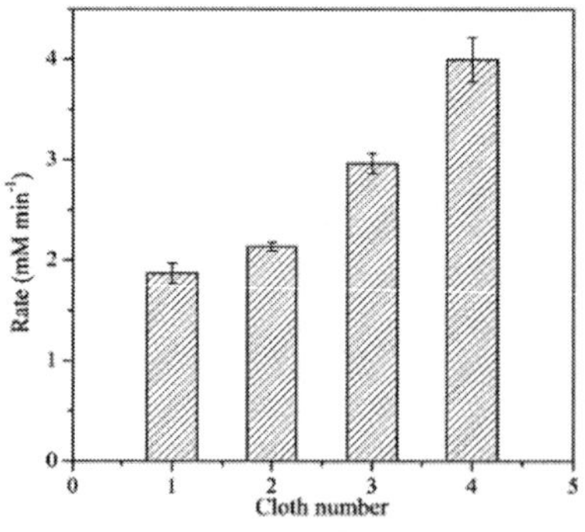

(b)

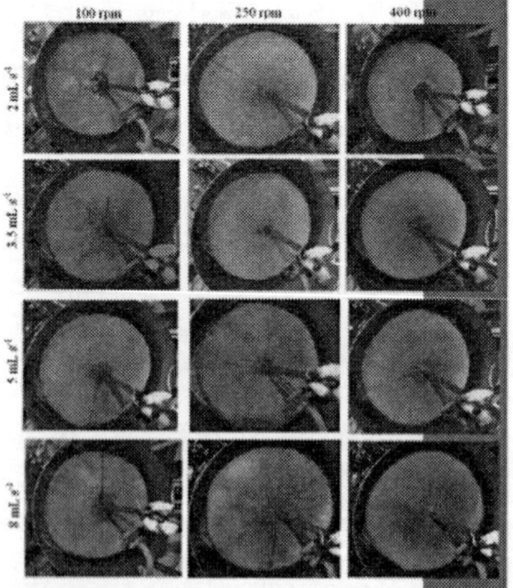

(c)

Spinning Mesh Disc Reactor

- ✓ Process Intensification
- ✓ High Reusability of enzyme
- ✓ Scale-up

Feng *et al.* (2013) *Chemical Engineering Journal*, 221, 407-417.

Towards a one pot cascade reaction

Dynamic Kinetic Resolution (DKR) of Henry Reaction

Perfecting individual reactions in the SMDR

STEP 1: Copper catalysed Henry reaction

STEP 2: Enzyme catalysed kinetic resolution of racemic alcohol

One-pot cascade reaction

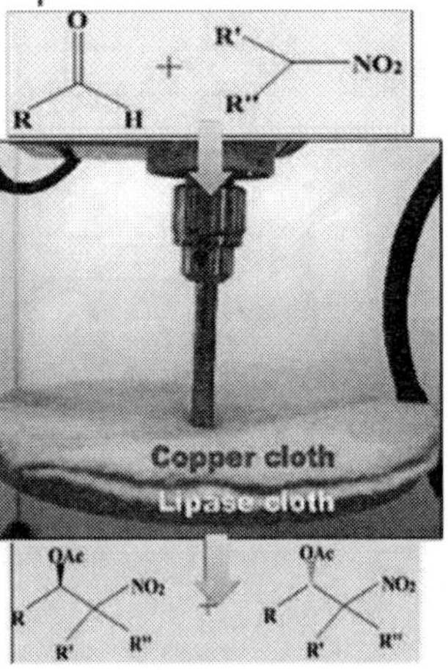

Copper-lipase catalyst in tandem

The Henry reaction

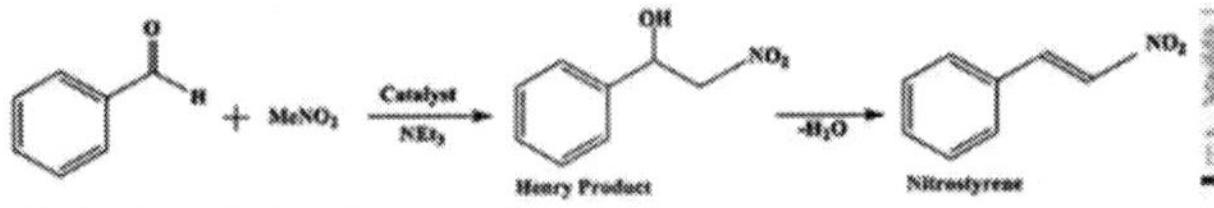

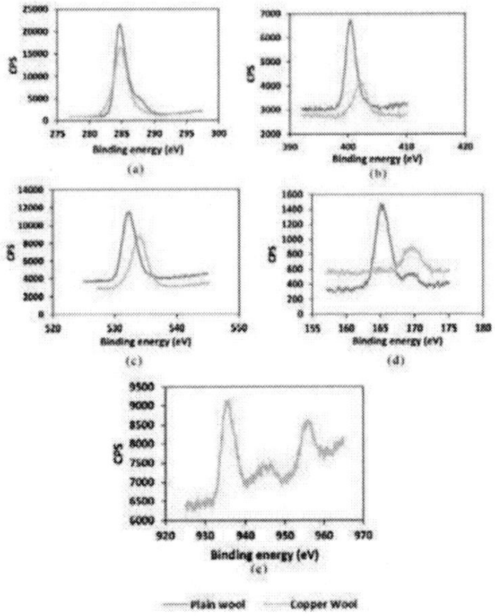

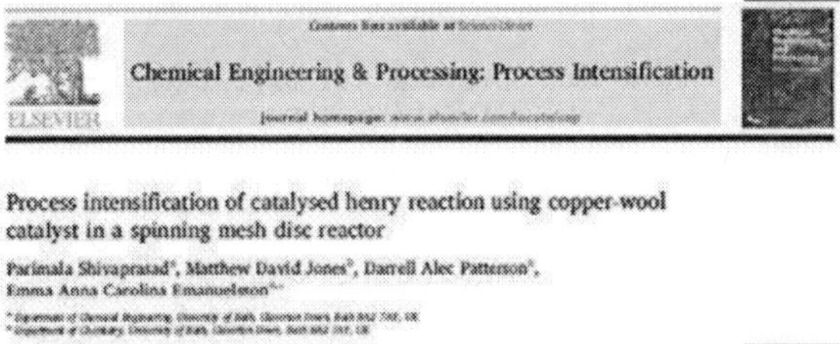
Chemical Engineering & Processing: Process Intensification

Process intensification of catalysed henry reaction using copper-wool catalyst in a spinning mesh disc reactor

Parimala Shivaprasad, Matthew David Jones, Darrell Alec Patterson, Emma Anna Carolina Emanuelsson

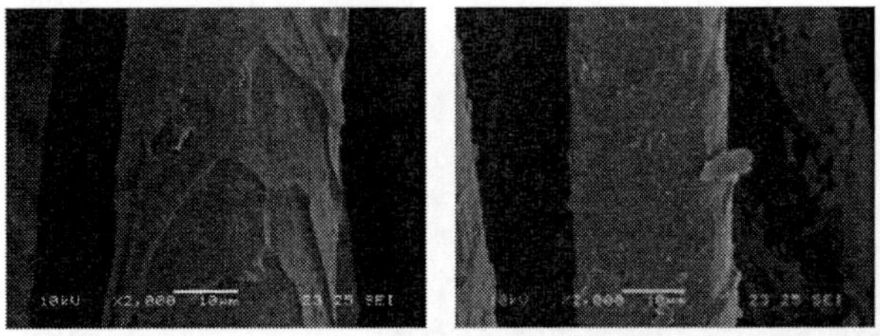

Copper triflate was successfully immobilised on wool

The Henry reaction

Henry reaction with free and immobilised catalyst in batch

Catalyst	Time (h)	% Conversion
Free copper triflate	4	38
(homogeneous batch)	6	42
	24	57
Copper cloth	4	41
(heterogeneous batch)	6	63
	24	85

- ✓ Copper immobilised on wool showed a higher conversion in batch.
- ✓ Higher reaction rate was observed in the SMDR using the copper cloth
- ✓ The SMDR was successfully applied for a solvent based reaction.

Henry reaction using the copper cloth in the SMDR

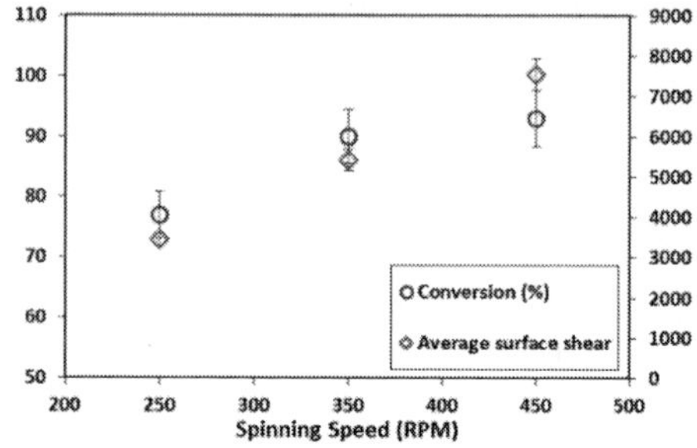

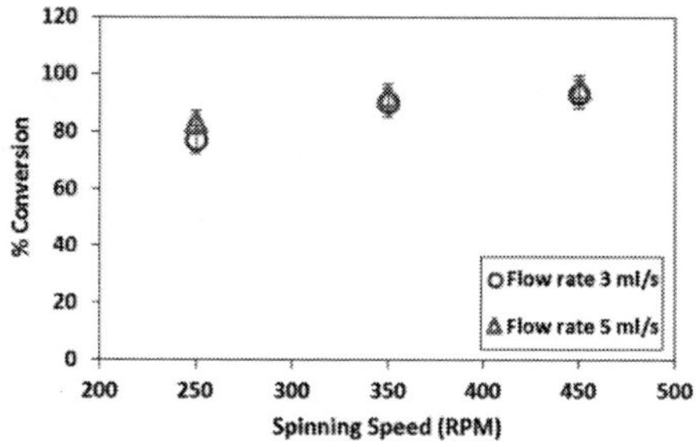

Shivaprasad, P., *et al.*(2017) *Chemical Engineering and Processing: Process Intensification.*

Wool as a catalyst

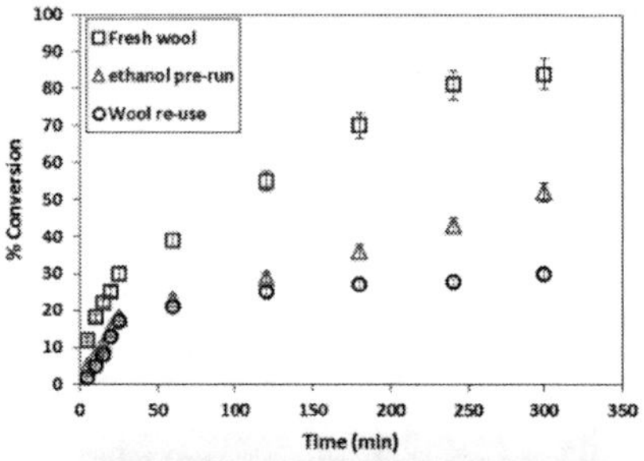

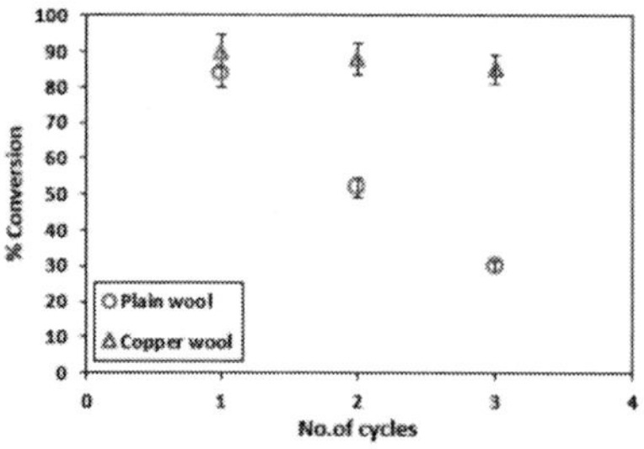

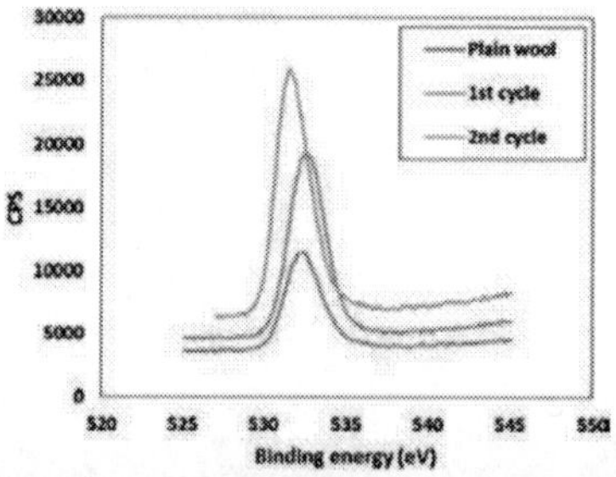

- ✓ Wool was found to catalyse the Henry reaction but was not re-usable for multiple cycles.
- ✓ Copper triflate immobilised on wool was more robust than plain wool.

Shivaprasad, P., *et al.*(2017) *Chemical Engineering and Processing: Process Intensification.*

Enzymatic kinetic resolution in the SMDR

UNIVERSITY OF BATH

Kinetic resolution of 1-phenyl ethanol catalysed by lipase

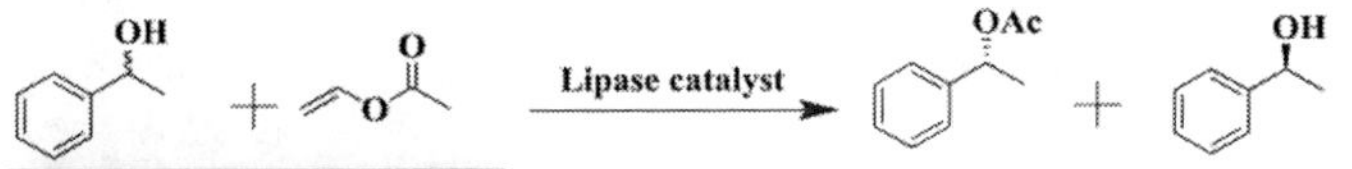

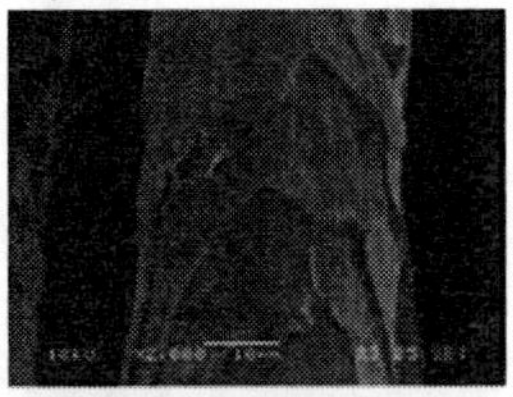

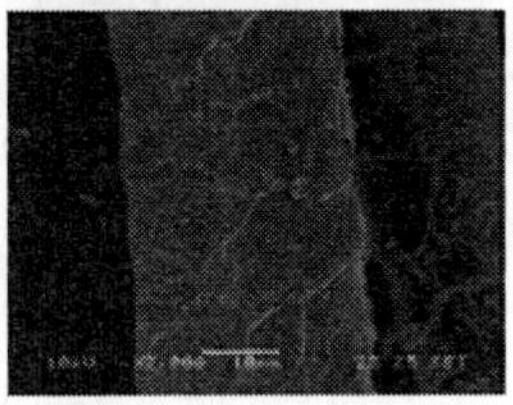

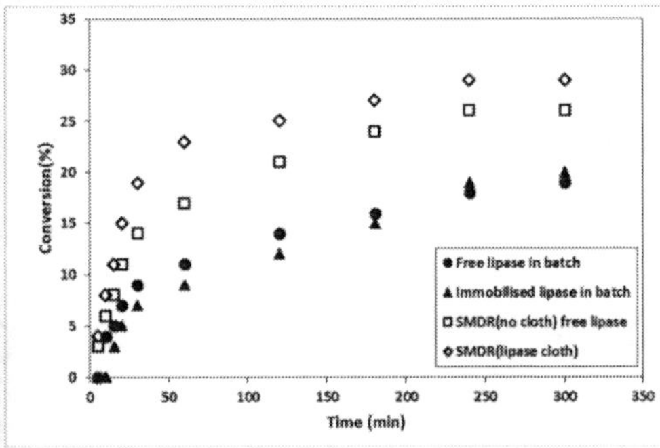

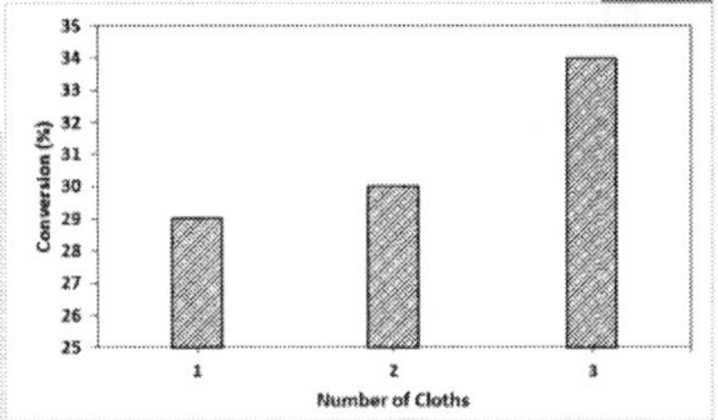

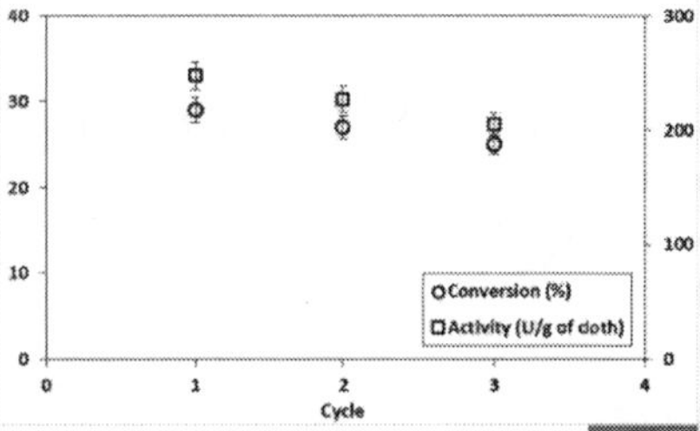

- ✓ Lipase immobilise on wool was used for the KR reaction in both batch and SMDR
- ✓ The reaction rate in the SMDR was higher for both free and immobilised lipase compared to the batch reaction.

TOWARDS CASCADE REACTIONS

UNIVERSITY OF BATH

Can copper and lipase immobilised on wool catalyse the reaction in tandem?

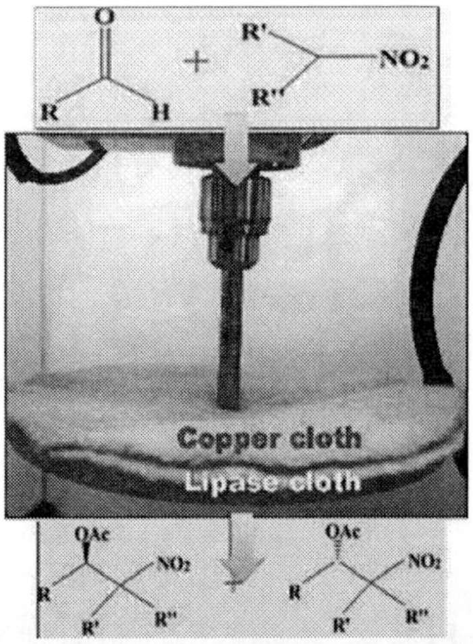

Challenges

- ✓ Solvent compatibility
- ✓ Robustness of the metal and the enzyme catalyst in one-pot
- ✓ Reaction kinetics

SUMMARY

UNIVERSITY OF BATH

Process intensification compared to more conventional reactors under equivalent operating conditions:

- enhanced reaction rate
- enhanced enzyme stability

The SMDR was used to demonstrate traditional organic synthesis reactions:

- The Henry reaction
- Enzyme catalysed kinetic resolution

Future work

UNIVERSITY OF BATH

One-Pot Cascade reaction

- Different enzyme and catalyst

Spinning Mesh Disc Reactor – further characterisation

- Increasing cloth size vs stacking cloths?
- A prototype for 'Perfect Reactor' – The future of process intensification

Acknowledgement

UNIVERSITY OF BATH

- Emanuelsson and Patterson Research Group
- Dr. Xudong Feng and Kemi Lawal
- The Santander mobility fund
- University of Bath and European Union's Seventh Framework Programme for research funding.

QUESTIONS?

484756 Scale and Effects of Catalyst Deactivation in Enzymatic Catalyzed Reactive Distillation

Abstract:

Increasing global competition and environmental legislation require the chemical sector to produce more cost efficient and reduce emissions. Process integration reduces the energy consumption of a process as well as its investment costs. An example for an integrated process is the combination of reaction and separation in form of the reactive distillation (RD) or reactive dividing wall column (RDWC). However, many of the industrially employed catalysts also enhance the production of unwanted side products that require additional, expensive purification steps. The employment of enzymatic catalysts is a promising approach to greatly increase the selectivity and reduce the required temperatures for chemical reactions. As enzymes are often product inhibited, the in situ product removal of a reactive distillation allows fully taking advantage of the enzymeâ€™s potential. Studies on laboratory and pilot plant scale columns have already shown the feasibility of this approach[1,2]. For the next step, the industrial application in a continuous reactive distillation, the stability of the enzymes is a fundamental condition, but yet untested under operating conditions. Therefore, knowledge about the scale of deactivation, the influencing factors and how to operate a column to minimize deactivation and maximize the production time before a standstill is required, are key questions still to be answered.

Our research aims to systematically investigate the behavior of eRDWCs. The first step has been to show the general feasibility of a successful operation with a continuous eRDWC 12 m high pilot plant.[1] Focus of this work was to develop a start-up strategy and to show that the column can be operated successfully for different operating conditions. Using the acquired experimental data a comprehensive simulation model was validated. The next step is a comprehensive approach to answer the questions about enzyme stability that hinder the industrial application. To analyze the scale and driving forces of enzyme deactivation two different experimental set-ups have been built. First, the eRDWC pilot plant and additionally a fixed-bed reactor have been employed in long term studies to test enzyme deactivation. After examining the experimental results on the quantity and driving forces of deactivation the second part of the project uses the developed mathematical model to analyze the influence on deactivation. The combined results from experimental and simulation studies allow a much deeper understanding about how to minimize enzyme deactivation in reactive distillation and additionally how to achieve the required conversion and purities for reduced enzyme activities.

This presentation will give a comprehensive overview about the scale and underlying causes of enzyme deactivation. A short introduction will explain how an eRDWC works and when to employ it. The simulation model and pilot plant will be presented. Selected experimental results about enzyme activity from the eRDWC and fixed-bed reactor experiments will be shown. From this data conclusions will be drawn about scale and influencing factors. Finally, it will be explained how to limit the enzyme deactivation by employing a smart process control and how to maintain the product specifications if deactivation occurs.

References

1.Egger T, Fieg G. Enzymatic catalyzed reactive dividing wall column: Experiments and model validation. AIChE J. 2016. doi:10.1002/aic.15598 .

2.Heils R, Jensen J, Wichert S, et al. Enzymatic Reactive Distillation: Kinetic Resolution of rac -2-Pentanol with Biocatalytic Coatings on Structured Packings. Ind. Eng. Chem. Res. 2015;54(38):9458-9467. doi:10.1021/acs.iecr.5b02802 .

498215 Use of Computational Fluid Dynamics (CFD) Simulation and Image Analysis Tool for Modelling Light in a Microalgal Photobioreactor

Abstract:

Light, one of the most important parameter for the growth of microalgal cells, is converted into chemical energy by the process of oxygenic photosynthesis. Microalga is a valuable feedstock for the production of industrially important metabolites. The energy for algal growth is obtained from the visible light within the photosynthetically active radiation (PAR) ranging from 400 to 700 nm wavelength. The high density of microalgal cultures is usually attained by cultivation in the closed photo-bioreactors (PBR) as compared to open systems due to better process control, mixing, light distribution etc. However, there is a need for optimizing the cultivation processes that should support both autotrophic and mixotrophic cultivation. Light in particular is known to directly influence the algal biomass production and product yield (Kumar, Sirasale, & Das, 2013). The conventional algal PBRs suffers from self-shading effect which is caused when microalga attains high cell density, thereby reducing the light penetration inside PBR and leading to development of dark zones. This is mainly because gradient of incident light intensity varies along the radius of PBR. Thus, local light intensity distribution at any given point inside the PBR is dependent on the path length, light penetration and absorption coefficient that becomes much acute especially during scaling-up of the process. Existing PBR designs do not have the desired efficiencies, and lead to high material and energy costs. Moreover, predicting the hydrodynamics (fluid flow pattern) of a PBR in laboratory experiments is difficult and cumbersome. Therefore, novel PBR designs need to be developed and benchmarked against existing designs.

Light utilization in the microalgal PBRâ€™s is usually measured by incident light intensity on the surface of the PBR or the irradiating surface area per unit culture volume. Average light intensity received by the microalgal cells is used for measuring the light utilization or photosynthetic efficiency of photosynthetic microorganisms. Efficient mixing allows cells to continuously pass through various zones inside a PBR such as complete dark, photo-limiting (cell growth increases with the light intensity), saturating and photo-inhibition (increase in light intensity leads to growth inhibition) and gives them flashing light effect. Despite several simulation studies on PBRs, a comprehensive analysis and evaluation of light penetration in PBRs has been limited.

The present study discusses use of CFD simulation for designing a novel airlift PBR with improved hydrodynamics of fluid flow. The novel PBR will be used for the cultivation of microalgae. The designed reactor performance in terms of light distribution will be compared with the conventional bubble column PBR. Since the application of multiphase CFD is complicated due to difficulty in describing the variety of interactions in these systems. The hydrodynamics of the reactor will be simulated using CFD and later will be used for microalgal cultivation. The present work describes the use of image analysis technique for the measurement and tracking of cell growth at different growth phases of a live bio-culture in the indigenously designed novel PBR. The light penetrability of incident light intensity was assessed in a Chlorella sp. culture throughout a batch experiment. The light penetration profile could be generated following modified Beer-Lambertâ€™s law by spectrophotometric method. Digital images of the top view of the PBRâ€™s were taken and processed using image processing tool in the MATLAB software. This was used to estimate the light intensity distribution in the externally radiating column PBR across the radial path length. Light gradient pattern inside the algal suspension was assessed by combining fluid dynamics simulations of three-dimensional turbulent single-phase fluid flow with statistical particle tracking and signal analysis. (Perner-Nochta & Posten, 2007). The experimental results and CFD simulations exhibited a good fit. Also, the light penetration was mapped with the image analysis tool for a better understanding of light availability and detection of dark zones inside a working PBR.

501946 Insights into the Relationship between Structure and Reactivity Descriptors in Realistic Nanoparticle Models: A DFT Approach

Abstract:

Fischerâ€“Tropsch (FT) synthesis is a promising road for the conversion of syngas into valuable hydrocarbons such as gasoline, diesel, chemicals and so forth. The chemical reactions are catalyzed more commonly by transition metals including Fe, Co and Ru. Albeit, only Fe and Co catalysts are commonly employed in industry. Due to the various steps involved during the conversion and the variety of products that can be generated, significant variation in catalyst composition and shape affects the nature and content of the products. That is, the catalytic activity of metal nanoparticles is greatly size-dependent and even just an additional atom can entirely change the reactivity and stability. Due to encouraging results with previous studies on 13-atom cluster models of Co, Fe, Ni, Pd, Pt, and Ru, we are now exploring 55-atom cluster models of the same metals using Density Functional Theory (DFT) calculations. These cluster models have shown to retain the accuracy of the periodic slab models at a lower computational cost. Since experimental information at the atomic level is difficult to obtain, we carried out systematic DFT calculations for evaluating reactivity descriptors of larger models. These cluster-based models offer a platform for the study of the adsorption and desorption of CO and H2 at the initial stages of the FT synthesis.

498860 Transportation Fuels from Renewable Hydrogen and Green Carbon Dioxide – a Technical, Economic, and Environmental Evaluation

Abstract:

The desire to reduce greenhouse gas emissions from the transportation sector along with the increasing penetration of fluctuating renewable power sources in several countries have led to an growing interest in the production of synthetic fuels from electricity [1]. Most of the corresponding processes employ water electrolysis in order to produce hydrogen. Apart from direct use in fuel cell electric vehicles, hydrogen can also be reacted with CO_2 to form various gaseous or liquid organic fuels that are more compatible with existing infrastructure and engines. Several pathways are being suggested in literature that use CO_2 from different sources [2] and lead to different fuels [3]. However, comprehensive comparisons to assess the benefits and drawbacks of these options including both fuel production and use in transportation are still scarce.

In this contribution, we present a detailed technical, environmental, and economic evaluation of several pathways that are considered promising. This includes direct application of hydrogen in a fuel cell vehicle, as well as conversion of hydrogen and CO_2 to combustion engine fuels like methane, methanol, and different methyl ethers, some of which can also be used as blend components with conventional fuels [4]. The CO_2 used is assumed to be obtained from biogas upgrading, where it occurs as a currently unused byproduct in high concentrations. Biogas upgrading is already being conducted at a number of operating biogas plants in order to obtain biomethane of higher purity [3]. The pathways are evaluated with respect to their well-to-wheel efficiency, greenhouse gas emissions, and fuel cost. A technical analysis is conducted based on detailed process simulations in order to highlight what causes the differences in the performance of the different pathways, and an outlook is given on the potential for improved production processes.

References:

[1] M. Bailera, P. Lisbona, L.M. Romeo, S. Espatolero. Renew. Sust. Energ. Rev. 69: 292–312, 2017.

[2] G. Reiter, J. Lindorfer. J. CO_2 Util. 10: 40–49, 2015.

[3] W. Wang, S. Wang, X. Ma, J. Gong. Chem. Soc. Rev. 40: 3703–3727, 2011.

[4] L. Lautenschütz, D. Oestreich, P. Seidenspinner, U. Arnold, E. Dinjus, J. Sauer. Fuel 173: 129–137, 2016.

492889 Hydraulic Retention Time and Temperature Impacts on Biogas Production in Expanded Granular Sludge Bed Reactor

Abstract:

Renewable sources play a significant role in providing sustainable energy with reduced Greenhouse Gas emissions considered to be a significant cause of global warming. Renewable energy resources combined in a hybrid energy system can also provide reduced energy costs. Anaerobic Digestion is one of these renewable energy sources, which involves the degradation of biomass in the absence of air to produce biogas that can be used in many applications. This technology also helps reduce the effluent waste discharged to a waste-water treatment system that further improves the eco-friendly nature of the hybrid system by lowering water loss to the environment.

An expanded granular sludge bed reactor, updated from the up-flow anaerobic sludge reactor, has been used to investigate the effect of hydraulic retention time and temperature on biogas production rate and methane composition of the biogas. The process consists of a two-stage reactor with a pre-acidification stage where hydrolysis and acidogenesis steps occur followed by the main reactor where the acetogenesis and methanogenesis steps occur.

Equivalent specific values of organic loading rates for industrial waste water with different COD levels (20, 30 and 40 g COD/l) have been used. The biogas production increases with decreasing hydraulic retention time (i.e. 24 gallons/day) and decreases when biogas production is inhibited by lower residence times below that required to complete degradation of the wastes and eventually affect the activity of the overall biodigestion process.

Methane composition and COD removal efficiency are (65-78%) and (90-96.5 %) respectively. A detailed analysis has been conducted for the process effluent using a Benchtop Spectrophotometer to monitor the process stability.

500041 Reverse Water Gas Shift Reaction over Cu-Fe/Al_2O_3 Catalyst in Solid Oxide Electrolysis Cells

Abstract:

Reverse water gas shift (RWGS) reaction is a practical method to convert CO_2 to CO utilizing hydrogen supplied from clean energy sources. Solid oxide electrolysis cells (SOECs) are new energy conversion devices with high efficiency and environmental compatibility. They can split steam (H_2O) into hydrogen and oxygen by using electricity generated from clean energy sources (such as solar energy and wind energy). Thus, applying the SOECs in the RWGS reaction could be a feasible way for efficient conversion of CO_2 using renewable energy sources. RWGS reaction was carried out over Fe-Cu/Al_2O_3 catalyst by a fixed-bed quartz tube reactor using externally supplied hydrogen at temperature of 700 °C, while fixing total flow rates of H_2 and CO_2 at 100 ml/min (CO_2:H_2 is 1:1). The catalyst showed good catalytic activity with CO
results confirmed that the sintering of active sites was prevented during the reaction. Cu-Fe/Al_2O_3 catalyst was then utilized as a functional reforming layer for the cathode of a conventional SOEC single cell. The single cell consisted of Ni-yttria stabilized zirconia (YSZ) cathode support/YSZ/$Ce_{0.8}Sm_{0.2}O_{1.9}$ (SDC) bi-layer electrolyte/$La_{0.6}Sr_{0.4}Co_{0.2}Fe_{0.8}O_{3-Î'}$ anode. Using this SOEC single cell with the catalysts layer, the RWGS reaction was investigated under the applied electric load condition. Our results indicated that by electrolysing H_2O over the cathode of SOEC cell, a reduced amount of external hydrogen is needed to obtain a similar CO_2 conversion as the fixed-bed quartz tube test.

494882 Electrochemical Hydrogenation of Bio-Oil Compounds Using a Polymer Electrolyte Membrane Reactor

Abstract:

Electrochemical hydrogenation (ECH) is an option for stabilizing or upgrading bio-oil that employs mild conditions (less than 80°C and ambient pressure). In this study, acetone, furfural and phenol are chosen as model substrates because of their abundance in bio-oil. The ECH method in a Polymer Electrolyte Membrane Fuel Cell (PEMFC) reactor has been considered to be one of the feasible ways of hydrogenating chemical compounds. In this study, acetone, furfural, and phenol which are present in pyrolysis bio-oil were studied for ECH. The main products from the ECH of acetone were isopropanol with selectivity approximately 90%. Diisopropyl ether was a byproduct. 16% propanol and 1% diisopropyl was identified in the final product using a new membrane, a holed bag collection method, the reaction temperature 65°C, and the reaction relative humidity of 90%. Propanol selectivity could be further improved by reducing the input power supply voltage.

The main challenge in ECH using a proton exchange membrane reactor is that a large fraction of current goes towards generating hydrogen gas that is not used in the hydrogenation reaction. High current efficiency (59.7%) could be achieved by using a PEMFC reactor. The mild operation condition is the greatest advantage of the proposed method, the optimized reaction condition is at 65 °C and 1 atmosphere. Relative humidity positively effects H_2 Utilization Efficiency and Current Efficiency, while temperature did not show linear influence during acetone ECH. The challenging part was the membrane contamination. Suitable reaction pressure, low reaction temperature, and low reaction relative humidity could prolong membrane life.

Parametric optimization study of downdraft biomass gasification using a comprehensive transport and kinetic model

Tapas Kumar Patra, Pratik N Sheth
Department of Chemical Engineering
BITS Pilani, Pilani Campus
INDIA

AIChE Annual Meeting 2017
October 29- November 3
Minneapolis Convention Center, Minneapolis, USA

17AIChE
Annual Meeting, Minneapolis, MN

Plan of Presentation

- Introduction
- Details of the developed model
  - Model Equations
  - Boundary conditions
  - Modeling Approach
- Operating Parameters
- Results and Discussion
- Conclusion
- References

Introduction

Why Biomass Energy ??

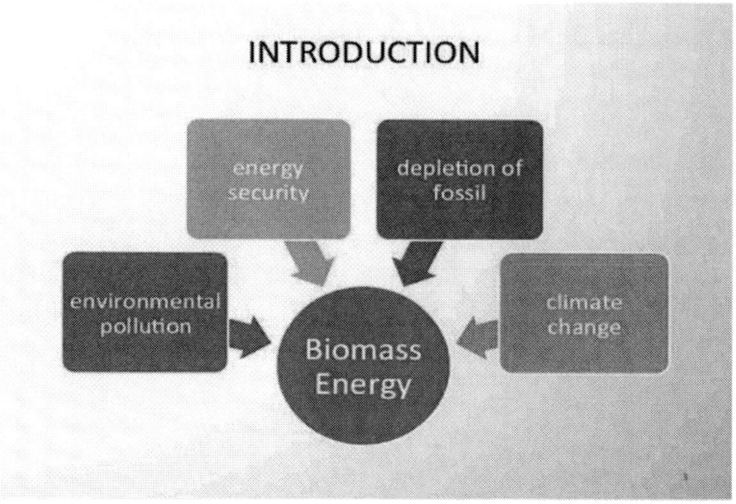

Downdraft Biomass Gasification

- Biomass gasification is the process converting biomass into gaseous fue
- Sub stoichiometric amount of air/oxygen is supplied
- In a downdraft gasifier both biomass and gasifying agent move in the downward direction
- Biomass passes through four zones
- Each zone performs different reactions and are distinguished based on temperature
- Boundaries of the zones are not rigid but transitional

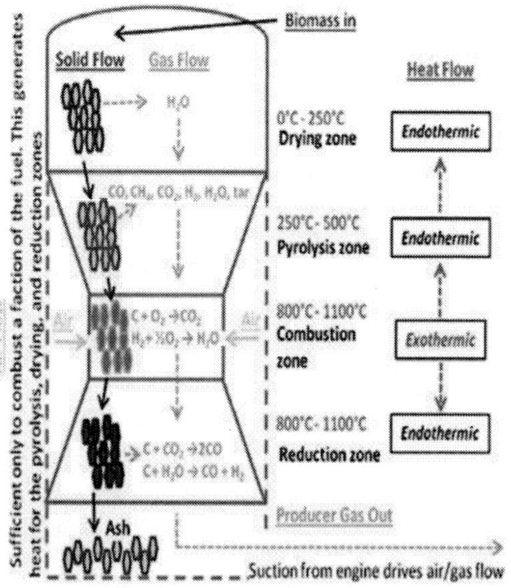

BITS Pilani, Deemed to be University under Section 3 of UGC Act, 1956

Gasification Reactions and Thermodynamics

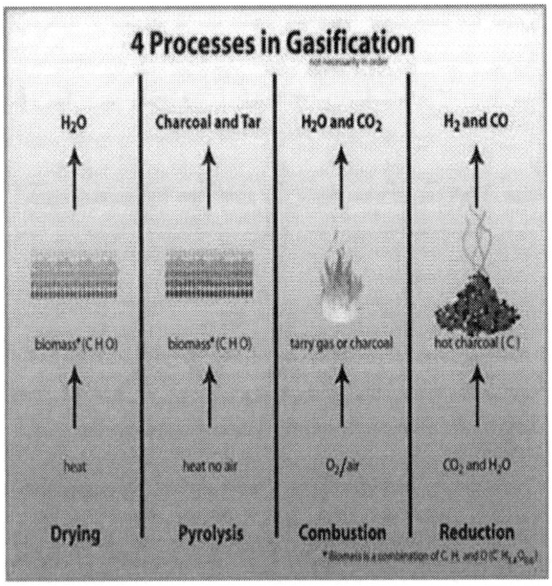

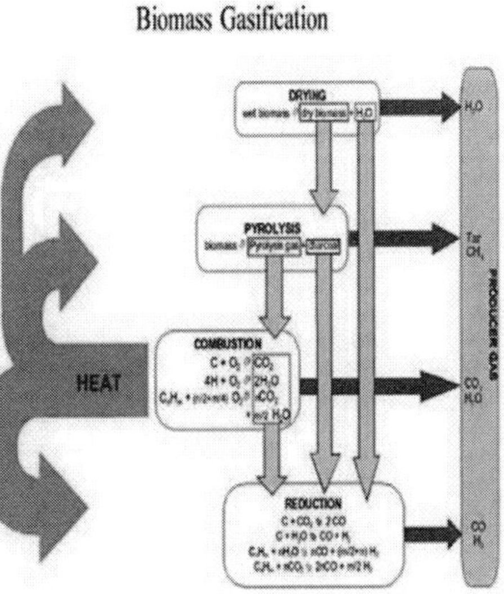

Overall Gasification Process

Preprocessing of biomass → Gasification → Gas clean-up & reforming → Gas utilization

Preprocessing of biomass	Gasification	Gas clean-up & reforming	Gas utilization
Size reduction, Drying, Preparation of gasifying agents	Heating, Chemical reactions, Catalysis	Cleaning of tar from syngas, Reforming of the syngas, Catalysis	Gas turbine, Gas burner, Fuel cell, Combined heat and power
Upstream processing	Gasification	Downstream processing	

Application of syngas

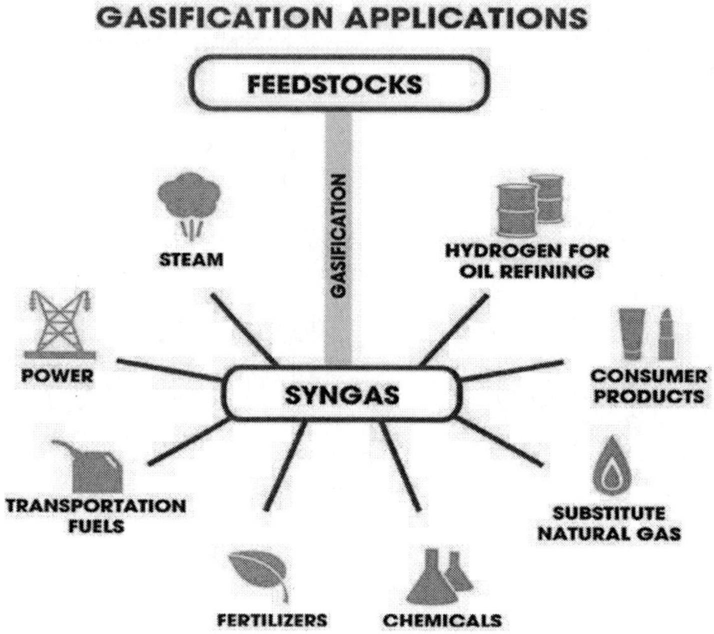

6

Gasification Process Simulation

innovate achieve lead

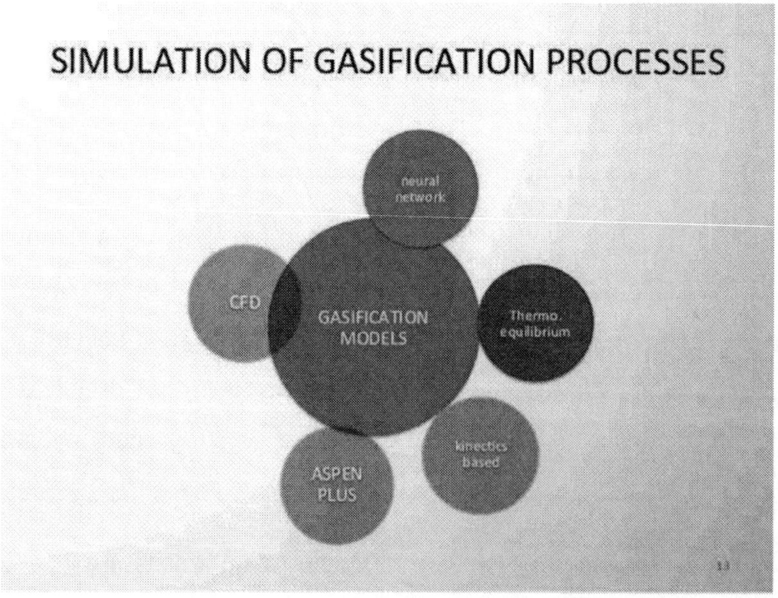

7

BITS Pilani, Deemed to be University under Section 3 of UGC Act, 1956

Schematic diagram of developed Model

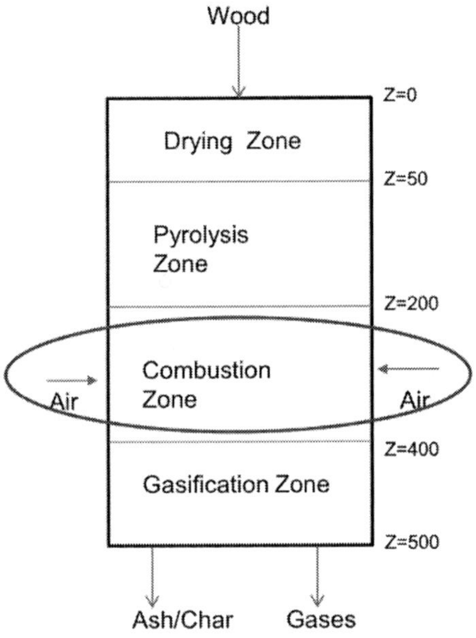

Model Equations

- **Wood**

$$\frac{\partial \rho_w}{\partial t} + \frac{\partial U_s \rho_w}{\partial z} = -r_{p1}$$

p1—primary pyrolysis

- **Moisture**

$$\frac{\partial \rho_M}{\partial t} + \frac{\partial U_s \rho_M}{\partial z} = -m_M$$

m_M-moisture evaporation/condensation

- **Oxygen**

$$\varepsilon \frac{\partial \rho_O}{\partial t} + \frac{\partial U_g \rho_O}{\partial z} = \frac{\partial}{\partial z}\left(\varepsilon D_O \rho_g \frac{\partial Y_O}{\partial z} \right) + M_O \sum_j v_{Oj} r_j + U_{gOi} \rho_{gO} Y_{OO} \delta(z - z_{Oi})$$

j=c1-c6

- **Gas Species**

$$\varepsilon \frac{\partial \rho_i}{\partial t} + \frac{\partial U_g \rho_i}{\partial z} = \frac{\partial}{\partial z}\left(\varepsilon D_i \rho_g \frac{\partial Y_i}{\partial z} \right) + M_i \sum_j v_{ij} r_j + v_i r_{p1} + v_i^* r_{p2}$$

i=H_2, CO, CO_2, CH_4, j=c1-c6,g1-g3,wg,sr1,sr2

- **Steam**

$$\varepsilon \frac{\partial \rho_{H_2O}}{\partial t} + \frac{\partial U_g \rho_{H_2O}}{\partial z} = \frac{\partial}{\partial z}\left(\varepsilon D_{H_2O} \rho_g \frac{\partial Y_{H_2O}}{\partial z} \right) + M_{H_2O} \sum_j v_{H_2Oj} r_j + U_{gOi} \rho_{gO} Y_{H_2OO} \delta(z - z_{Oi}) + m_M + v_{H_2O} r_{p1} + v_{H_2O} r_{p2}$$

j=c1-c6,g1-g3,wg,sr1,sr2

- **Primary Tar**

$$\varepsilon \frac{\partial \rho_{T1}}{\partial t} + \frac{\partial U_g \rho_{T1}}{\partial z} = \frac{\partial}{\partial z}\left(\varepsilon D_{T1} \rho_g \frac{\partial Y_{T1}}{\partial z} \right) - M_{T1} r_{c1} + v_{T1} r_{p1} - r_{p2}$$

- **Nitrogen**

$$\rho_{N_2} = \frac{P M_{N2}}{RT} \qquad \rho_g = \Sigma \rho_i$$

- **Total Gas Continuity**

$$\varepsilon \frac{\partial \rho_g}{\partial t} + \frac{\partial (U_g \rho_g)}{\partial z} = + \sum_i \sum_j v_{ij} M_i r_j + m_M + (1 - v_C) r_{p1} + U_{gOi} \rho_{gO} \delta(z - z_{Oi})$$

i=O_2,H_2,CO,CO_2,H_2O,CH_4; j=c6,g1-g3

Model Equations

- **Solid Phase Enthalpy**

$$\frac{\partial(\sum_i \rho_i H_{si})}{\partial t} = \frac{\partial}{\partial z}\left(\lambda_s^* \frac{\partial T_s}{\partial z}\right) + \frac{\partial(U_s \sum_i \rho_i H_{si})}{\partial z} - \sum_j r_j \Delta H_j - Q_{sg} + Q_{sw} - m_M \Lambda$$

$$H_{si} = C_{si}(T_s - T_o), \quad i = W, M, C$$

- **Gas Phase Enthalpy**

$$\varepsilon \frac{\partial(\sum_i \rho_i H_{gi})}{\partial t} = \frac{\partial}{\partial z}\left(\lambda_g^* \frac{\partial T_g}{\partial z}\right) + \frac{\partial\left(U_g \sum_i \rho_i H_{gi}\right)}{\partial z} - \sum_j r_j \Delta H_j + Q_{sg} + Q_{gw} + U_{g0i}\rho_{g0}C_{g0}T_{g0}\delta(z - z_0)$$

$$H_{gi} = C_{gi}\left(T_g - T_o\right), \quad i = N_2, CO_2, CO, H_2, O_2, CH_4, H_2O, T$$

Solid-wall heat transfer

$$Q_{sw} = \frac{4h_{sw}}{D}(T_w - T_s)$$

Gas-wall heat transfer

$$Q_{gw} = \frac{4h_{gw}}{D}(T_w - T_g)$$

Solid-gas heat transfer

$$Q_{sg} = h_{sg}A_p \upsilon_p (T_s - T_g)$$

$$A_p \upsilon_p = \frac{6(1 - \varepsilon)}{d_p}$$

Boundary Conditions

- At the top of the reactor (z=0), temperature, velocity and density of the inlet stream is assumed constant and the solid is assumed to be at the ambient temperature T_e

  ρ_{wood}= 668.0 kg/m^3

  $\rho_{moisture}$= 66.8 kg/m^3 *(Considering 10% Moisture content)*

  $\rho_{H_2} = \rho_{CO} = \rho_{CO_2} = \rho_{CH_4} = \rho_{H_2O} = \rho_{T_1} = 0.0$

  $T_s = T_e$= 300.0 K

At the air inlet,

  ρ_{O_2} = 0.248619kg/m^3*(0.21 * air density at 25 C temperature (1.1839))*

  ρ_{N_2}= 0.935281kg/m^3 *(0.79 * air density at 25 C temperature (1.1839))*

  ρ_g= 1.1839 kg/m^3

  $Ug = U_{g0}$=0.2m/s

- At the bottom of the reactor(Z=L)

$$\frac{\partial T_s}{\partial z} = \frac{\partial T_g}{\partial z} = \frac{\partial \rho_i}{\partial z} = 0, (i = CO_2, CO, H_2, CH_4, H_2O, T, N_2, O_2)$$

Modeling Approach

- The reactor is divided into a set of elementary cylindrical cells
- Finite Difference Method (FDM) was applied for discretization
- After discretization, the equations were solved with Semi – implicit procedure i.e. each conservation equation is implicit in the corresponding variable being conserved whereas the other variables are taken as the last available values.
- Time step = 10^{-5}s
- Space step=0.01m
- Grid length = length/space step
- All the constant values were taken as stated in the literature

Model Validation and Parametric Study

- The model is validated with the experimental data of Sheth *et al.* (2010)
- The model has been validated for experimental parameters such as
  - Producer gas flow rate
  - Biomass consumption rate
  - Char production rate
- Parameters chosen for the study are
  - Air flow rate
  - Biomass moisture content
  - Reduction zone length

Results: Air Flow Rate

- Air flow rate is one of the vital parameters for gasification process
- Sub-stoichiometric amount of air is supplied to avoid complete combustion
- Increase in air flow rate dilutes the product gas since nitrogen dominates in the composition
- From figure it is seen that the composition of CO and H_2 decreases continuously and for N_2 and CO_2 it increases.
- It clearly shows that combustion dominates with the increase in air flow rate

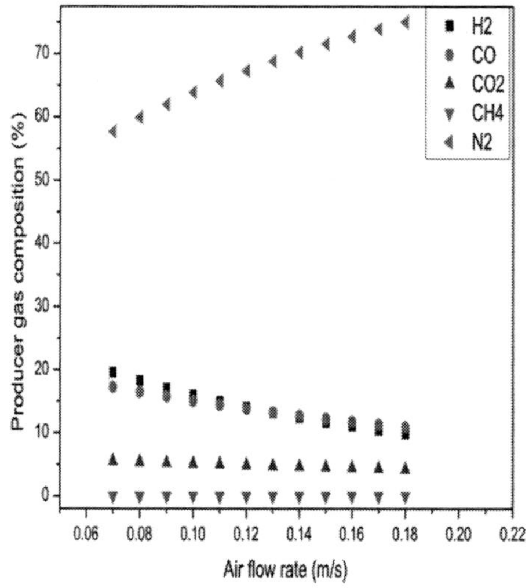

Model predicted product gas composition after 30 minutes for different air flow rates

Results: Air Flow Rate

$$Equivence\ ratio = \frac{\left(Air/_{Fuel}\right)_{actual}}{\left(Air/_{Fuel}\right)_{Stoichiomeric}}$$

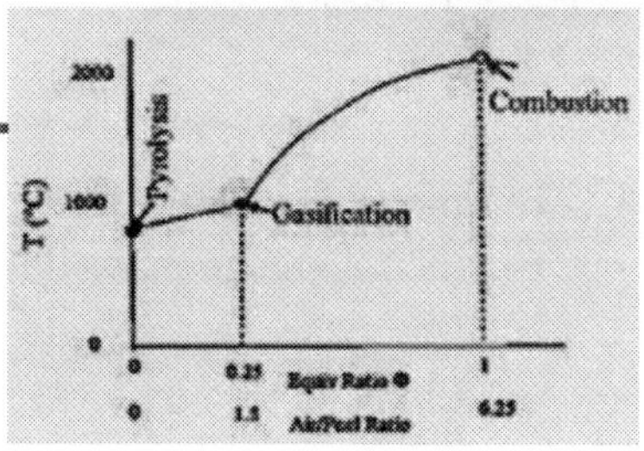

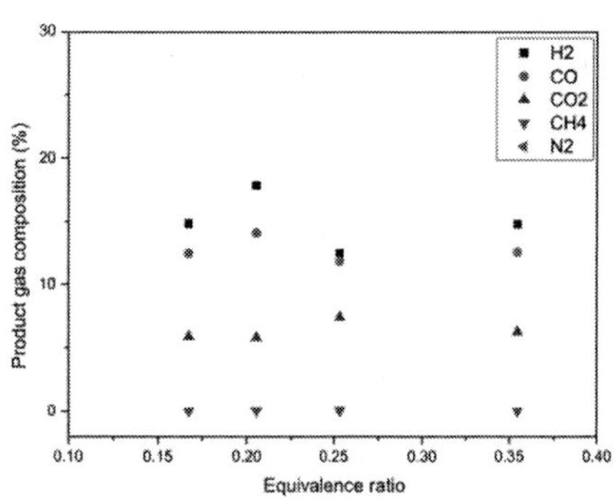

Model predicted product gas composition for different ER

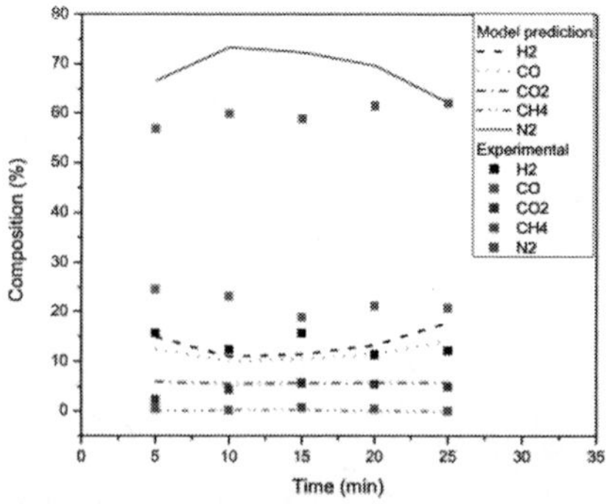

Comparison of Model predicted product gas composition with experimental data for ER 0.2054

BITS Pilani, Deemed to be University under Section 3 of UGC Act, 1956

Results: Moisture Content

- Moisture content of biomass decides the composition of syngas
- Suitability of a biomass depends on the gasification process
- For downdraft biomass gasification the moisture content range varies between 5 % to 18 % wet basis
- Higher moisture content leads to deterioration of product gas quality

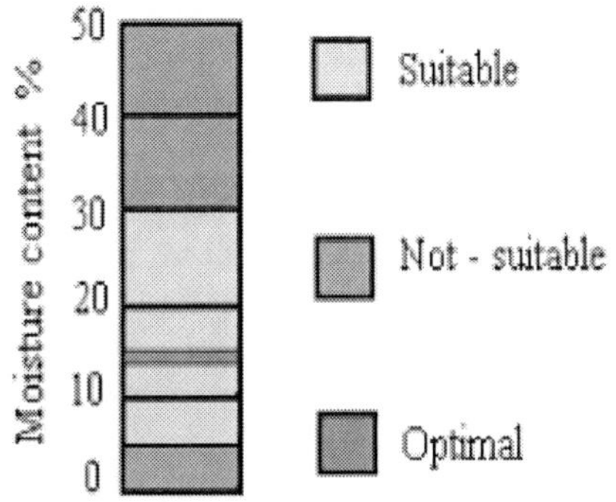

Moisture Scale for Gasification

Results: Moisture Content

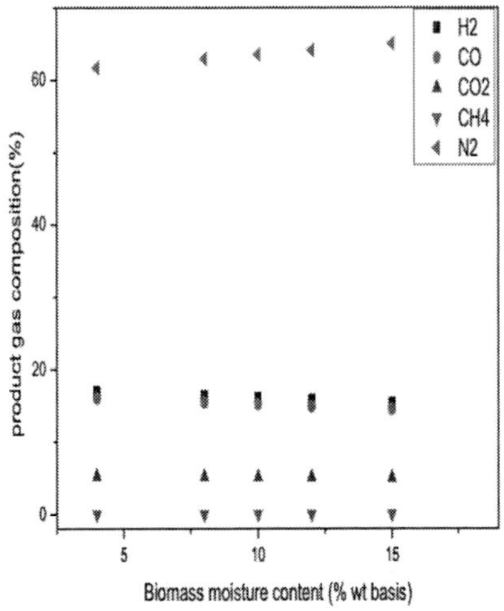

Model predicted product gas composition for different moisture content

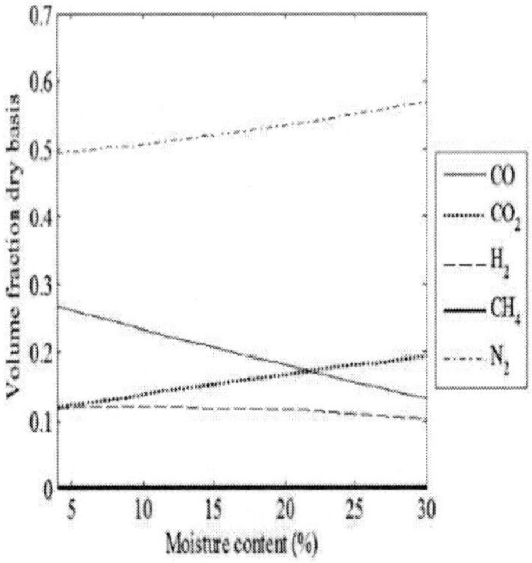

product gas composition for different moisture content reported by Atnaw *et al.* (2017)

ATNAW, S., SULAIMAN, S., SINGH, L., WAHID, Z., KU MOHAMMAD FAIZAL BIN CHE KU YAHYA, C.. Modeling and Parametric Study for Maximizing Heating Value of Gasification Syngas. **BioResources**, North America, 12, feb. 2017

Results: Reduction zone length

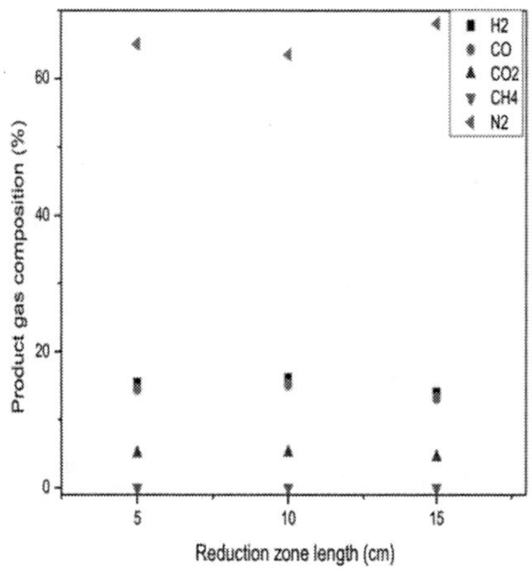

Model predicted product gas composition for different reduction zone lengths

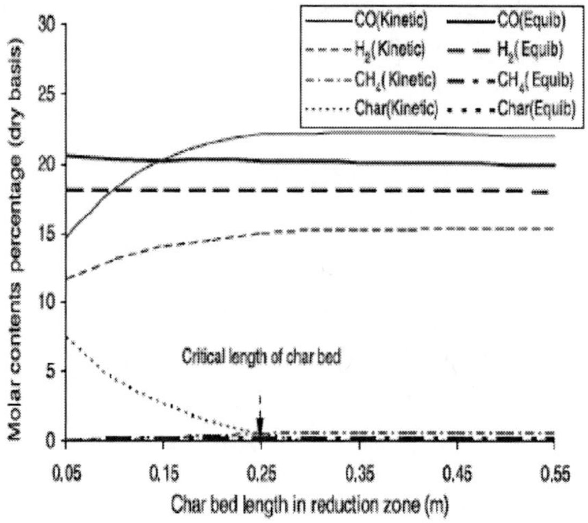

Model predicted product gas composition for different char bed lengths reported by Sharma (2008)

Sharma AK. Equilibrium and kinetic modeling of char reduction reactions in a downdraft biomass gasifier: A comparison. Solar Energy 2008;82(10):918-28.

Conclusions

- A comprehensive transport and kinetic model for downdraft biomass gasification has been developed
- The model has been validated with experimental data reported by Sheth *et al.* (2010).
- A parametric study using Air flow rate, biomass moisture content and reduction zone length has been carried out.
- The increase in air flow rate from 0.08 to 0.18 m/s results in decrease of CO and H_2 composition in the product gas and increase of N_2 composition
- With the increase in biomass moisture content from 4% to 15% (wet basis) CO composition decreases and N_2 composition increases
- Reduction zone length variation from 5 cm to 10 cm resulted in increase in CO and H_2 composition, however at 15 cm the composition of CO and H_2 decreased

BITS Pilani, Deemed to be University under Section 3 of UGC Act, 1956

Funding

I would like to thank Department of Science and Technology (DST) Govt. of India for funding me to attend and present my paper at AIChE Annual Meeting 2017 Minneapolis, MN, USA.

innovate achieve lead

THANK YOU

490602 Hybrid Geothermal Energy Conversion - a Potential Solution for Low-Temperature Geothermal Resources

Abstract:

Low- to moderate- temperature geothermal resources are typically used for direct heat energy applications, whereas deep, high-temperature, geothermal resources are often used for electricity generation. But deep geothermal resources usually require enhancement of the permeability of the hot dry rock at these depths (4-6 km), which can induce seismicity. Here we present an alternative geothermal energy utilization approach that uses shallower, lower-temperature, naturally permeable formations. Our approach can generate electricity while substantially reducing drilling costs and the risk of induced seismicity. This approach uses geothermal heat to supplement a secondary energy source, thereby potentially enabling the use of geothermal energy for electricity generation in many regions worldwide that are not viable with current approaches. We determine the net power output, energy conversion efficiencies, and economics of these hybrid power plants, where the geothermal power plant uses supercritical carbon dioxide ($scCO_2$) as the subsurface working fluid. When $scCO_2$ is used underground, the overall system stores 100% of the injected CO_2, while generating geothermal power, constituting a Carbon Capture Utilization and Storage (CCUS) system. Parameters varied include geothermal reservoir depth (2.5-4.5 km) and turbine inlet temperature (100-220°C). We find that hybrid power plants outperform two individual, i.e., stand-alone geothermal and waste-heat power plants. Furthermore, such hybrid power plants are more economical, and result in lower capital costs per electricity produced ($/kW) than separate power plants.

497836 On the Origin of Preferred Bicarbonate Production from Carbon Dioxide (CO_2) Capture into Aqueous 2-Amino-2-Methyl-1-Propanol (AMP)

Abstract:

Chemical absorption by aqueous amine-based solvents appears the most promising near-term solution for CO_2 capture from flue gas because of their ability to capture CO_2 at low pressure with adequate absorption/desorption kinetics.[1,2] Monoethanolamine (MEA) is one of the most widely used and studied alkanolamines, but its expansion for commercial use tends to be limited due to degradation problems and high parasitic energy consumption during solvent regeneration.[3-4] 2-amino-2-methyl-1-propanol (AMP) and other sterically hindered primary amines have been suggested as an alternative because they exhibit both high CO_2 loading capacities and relatively fast absorption rates, in addition to low degradation and corrosion rates.[5-8] However, the mechanism underlying the reaction between CO_2 and aqueous AMP still remains controversial.

It is now well adopted that CO_2 capture by most primary amines including MEA in an aqueous solution takes place via a two-step mechanism involving a zwitterion intermediate.[9-10] The zwitterion may subsequently undergo deprotonation by another amine to form a carbamate ion and a protonated amine, leading to a loading capacity of 0.5 mol CO_2/mol amine. Conversely, the primary products of the reaction of aqueous AMP with CO_2 have been found to be bicarbonate (HCO_3^-) and protonated AMP ($AMPH^+$), with carbamate ($AMPCOO^-$) but whose concentration is about an order of magnitude lower than HCO_3^-.[11] It is also known that bicarbonate formation predominantly takes place in aqueous tertiary amines such as methyldiethanolamine (MDEA), yielding the theoretical maximum capacity of 1 mol CO_2/mol amine.[12] In the tertiary amine system, bicarbonate can be formed by the base-catalyzed hydration of CO_2, e.g., MDEA + H_2O + CO_2 → $MDEAH^+$ + OH^- + CO_2 ® $MDEAH^+$ + HCO_3^-.[13]

However, according to previous experimental observations, the absorption rate of CO_2 in aqueous AMP tends to be about two orders of magnitude faster as compared to the MDEA case.[14] On the basis of the much higher absorption rate of AMP relative to tertiary amines, it has been speculated that first AMP and CO_2 react to form carbamate ($AMPCOO^-$), which undergoes hydrolysis to form bicarbonate due to its possible instability (i.e., $AMPCOO^-$ + H_2O → HCO_3^- + $AMPH^+$).[5,15] However, previous theoretical studies have predicted free energy barriers for the hydrolysis of both $AMPCOO^-$ and $MEACOO^-$ to HCO_3^- to be high and also comparable, implying that not only $MEACOO^-$ but also $AMPCOO^-$ are relatively stable and may not easily undergo hydrolysis to form HCO^{3-}.[16–19] Rather, it would be more likely that the carbamates would revert back to the amine and CO_2.[20] This gives a hint that preferred bicarbonate formation in the AMP-H_2O-CO_2 system might not be due to carbamate being less stable than the MEA case.

In this work, we investigate the factors underlying the preferred production of bicarbonate over carbamate from CO_2 absorption in aqueous AMP, with comparison to the MEA case where carbamate is predominantly formed, by explicitly taking into account both thermodynamic and kinetic contributions. First, we evaluated the thermodynamic favorability for bicarbonate and carbamate formation by calculating the changes in free energy (and enthalpy) using static QM with an implicit solvent model (and AIMD with explicit solvent). Our calculations show that both CO_2 capture mechanisms are thermodynamically favorable with similar exothermicities in aqueous AMP and MEA solutions. In addition, AIMD simulations demonstrate that AMP carbamate ($AMPCOO^-$) can be as stable as MEA carbamate ($MEACOO^-$) in an aqueous solution, which is consistent with previous theoretical studies predicting large and similar energy barriers for carbamate hydrolysis to form bicarbonate in both cases[19,17], i.e., $AMPCOO^-$ ($MEACOO^-$) + $H_2O \rightarrow HCO_3^-$ + $AMPH^+$ ($MEAH^+$); the comparable stability between $AMPCOO^-$ and $MEACOO^-$ is also well demonstrated by analysis of the N-C (in CO_2) bonding interaction. This may suggest that the relative concentrations of carbamate and bicarbonate could not be predicted only by their thermodynamic favorability, especially since it may take too long to reach equilibrium at typical operating conditions.

From AIMD simulations, we observed that the AMP + $H_2O \rightarrow AMPH^+ + OH^-$ reaction frequently occurs, rather than carbamate formation (i.e., 2AMP + $CO_2 \rightarrow AMPCOO^- + AMPH^+$) which tends to occur readily in the case of MEA. If OH^- does not abstract a proton to form H_2O, bicarbonate can be formed via the OH^- + $CO_2 \rightarrow HCO_3^-$ reaction.[21,22] This can also occur in one step, via the amine-catalyzed hydration of CO_2 proposed by Donaldson and Nguyen[13], (i.e., AMP + H_2O + $CO_2 \rightarrow AMPH^+ + HCO_3^-$). These results suggest that bicarbonate formation may be more kinetically favorable relative to carbamate formation in aqueous AMP. The enhanced AMP protonation reaction is apparently related to its high basicity. Our electronic structure analysis shows that the steric hindrance of the CH_3 groups of AMP causes a more planar-like configuration around the central C atom (C_N) which is attached to N. The tendency of sp^2-like hybridization results in charge redistribution in which the C_N atom is more positively charged and the N atom is more negatively charged. This in turn strengthens the interaction between the N in AMP and surrounding H_2O molecules while suppressing the accessibility of CO_2 to the N site to form carbamate, relative to the MEA case, as confirmed by analysis of radial/spatial distribution functions from MD simulations.

Based on our calculation results, we attribute the preferential formation of bicarbonate in CO_2 absorption into an aqueous AMP solution largely to kinetic factors. As discussed above, the strong interaction between N (in AMP) and H (in H_2O) suppresses the reaction with CO_2 and promotes the protonation reaction, while the bicarbonate and carbamate reaction routes exhibit similar thermodynamic favorability. This study highlights that not only thermodynamic but also kinetic factors should be considered in describing CO_2 capture by aqueous amines.

References

1. Rochelle GT. Amine scrubbing for CO_2 capture. *Science*. 2009;325(5948):1652-4.

2. Figueroa JD, Fout T, Plasynski S, McIlvried H, Srivastava RD. Advances in CO_2 capture technology—The U.S. Department of Energy's Carbon Sequestration Program. *Int J Greenh Gas Control*. 2008;2(1):9-20.

3. Zhu D, Fang M, Lv Z, Wang Z, Luo Z. Selection of Blended Solvents for CO_2 Absorption from Coal-Fired

Flue Gas. Part 1: Monoethanolamine (MEA)-Based Solvents. *Energy & Fuels*. 2012;26:147-153.

4. Goff GS, Rochelle GT. Monoethanolamine Degradation: O_2 Mass Transfer Effects under CO_2 Capture Conditions. *Ind Eng Chem Res*. 2004;43(20):6400-6408.

5. Sartori G, Savage DW. Sterically Hindered Amines for CO_2 Removal from Gases. *Ind Eng Chem Fundam*. 1983;22(2):239-249.

6. Chakraborty AK, Astarita G, Bischoff KB. CO_2 Absorption in Aqueous Solutions of Hindered Amines. *Chem Eng Sci*. 1986;41(4):997-1003.

7. Lepaumier H, Picq D, Carrette P-L. New Amines for CO_2 Capture. I. Mechanisms of Amine Degradation in the Presence of CO2. *Ind Eng Chem Res*. 2009;48(20):9061-9067.

8. Lepaumier H, Picq D, Carrette P-L. New Amines for CO_2 Capture. II. Oxidative Degradation Mechanisms. *Ind Eng Chem Res*. 2009;48(20):9068-9075.

9. Caplow M. Kinetics of Carbamate Formation and Breakdown. *J Am Chem Soc*. 1968;90(24):6795-6803.

10. Danckwerts PV. The Reaction of CO_2 with Ethanolamines. *Chem Eng Sci*. 1979;34(4):443-446.

11. Ciftja AF, Hartono A, Svendsen HF. Experimental study on carbamate formation in the AMP–CO_2–H_2O system at different temperatures. *Chem Eng Sci*. 2014;107:317-327.

12. Jakobsen JP, Krane J, Svendsen HF. Liquid-Phase Composition Determination in CO_2 - H_2O - Alkanolamine Systems: An NMR Study. *Ind Eng Chem Res*. 2005;44(26):9894-9903.

13. Donaldson TL, Nguyen YN. Carbon Dioxide Reaction Kinetics and Transport in Aqueous Amine Membranes. *Ind Eng Chem Fundam*. 1980;19(3):260-266.

14. Vaidya PD, Kenig EY. CO_2-Alkanolamine Reaction Kinetics: A Review of Recent Studies. *Chem Eng Technol*. 2007;30(11):1467-1474.

15. Alper E. Reaction Mechanism and Kinetics of Aqueous Solutions of 2-Amino-2-methyl-1 -propanol and Carbon Dioxide. *Ind Eng Chem Res*. 1990;29(8):1725-1728.

16. Xie H-B, He N, Song Z, Chen J, Li X. Theoretical Investigation on the Different Reaction Mechanisms of Aqueous 2-Amino-2-methyl-1-propanol and Monoethanolamine with CO_2. *Ind Eng Chem Res*. 2014;53(8):3363-3372.

17. Davran-Candan T. DFT Modeling of CO_2 Interaction with Various Aqueous Amine Structures. *J Phys Chem A*. 2014;118(25):4582-4590.

18. Mehdizadeh H, Gupta M, Kim I, Da Silva EF, Haug-Warberg T, Svendsen HF. AMP–CO_2–water thermodynamics, a combination of UNIQUAC model, computational chemistry and experimental data. *Int J Greenh Gas Control*. 2013;18:173-182.

19. Yamada H, Matsuzaki Y, Higashii T, Kazama S. Density Functional Theory Study on Carbon Dioxide Absorption into Aqueous Solutions of 2-Amino-2-methyl-1-propanol Using a Continuum Solvation Model. *J Phys Chem A*. 2011;115(14):3079-86.

20. Da Silva EF, Svendsen HF. Computational chemistry study of reactions, equilibrium and kinetics of chemical CO_2 absorption. *Int J Greenh Gas Control*. 2007;1(2):151-157.

21. Stowe HM, Vilčiauskas L, Paek E, Hwang GS. On the origin of preferred bicarbonate production from

carbon dioxide (CO_2) capture in aqueous 2-amino-2-methyl-1-propanol (AMP). *Phys Chem Chem Phys.* 2015;17(43):29184-29192.

22. Nakai H, Nishimura Y, Kaiho T, Kubota T, Sato H. Contrasting mechanisms for CO2 absorption and regeneration processes in aqueous amine solutions: Insights from density-functional tight-binding molecular dynamics simulations. *Chem Phys Lett*. 2016;647:127-131.

493040 Evaluating the Performances of Reduction Process in the Solar Thermochemical Two-Step CO_2 Splitting Based on Ceria Redox Reactions

Abstract:

The emission of carbon dioxide is a severe problem in the current society. There are tremendous technologies that provide approaches to address the surplus CO_2. Solar thermochemical reactor provides an attractive approach that utilizes the most common solar radiation as the thermal driving force to motivate the reaction between CO_2 and metal oxides, which is so called metal oxide redox pair based thermochemical cycles. By two-step redox process, the incoming CO_2 can be transformed to CO for further applications. Among the diversified metal oxide redox pairs, $CeO_2/CeO_{2-\text{î}}$· shows an anticipated performance, due to its fast redox kinetics, high crystallographic stability of a wide range of reacting oxygen non-stoichiometry, and relatively high oxygen solid-state conductivity.

In this work, a numerical analysis is applied to study on the performance of CeO_2 reduction process in a segment of cavity-receiver reactor. The CeO_2 is presented as catalyst particles in the domain to mimic the structure of porous medium. A three-dimension transient model is applied to simulate the momentum, heat, and mass transfer in the domain. CeO_2 particle size, void fraction of domain, purge gas velocity and radiation flux rate are four key parameters that are analyzed in this work. By investigating on the influences of each factor to the reduction reaction, an optimized reaction rate can be achieved, which is beneficial to improve the solar to fuel efficiency of the reactor.

494372 Mass Transfer Studies of Carbondioxide Absorption in Sodium Hydroxide in Millichannels

Abstract:

During the past several decades, CO_2 levels in the atmosphere have been dramatically increasing due to the man-made emissions such as fossil fuel-fired power plants. This results in global warming thus leading an increase in the need to study different ways to capture the emitted CO_2 directly from the exhausts of power plants or atmosphere. There are several ways to remove CO_2, such as absorption into a liquid solvent, adsorption into a solid, cryogenic separation, permeation through membranes and photochemical conversion. In most industries, the absorption of CO_2 in chemical solvents (in absorption towers) is used for CO_2 capture. In these towers, the mass transfer along with chemical reactions take place between the gas and liquid phase. It is important to understand these processes in detail. These flow patterns are difficult to maintain in large scale industrial absorbers. So to get accurate information controlled gas-liquid absorption experiments are carried out in milli-channels in this work under controlled atmosphere. The absorption experiments of CO_2 in varying concentrations of sodium hydroxide solution are carried out in T-junction glass milli-channels with a circular cross section (inner diameter of 2mm). The gas and liquid flow rates are controlled by a mass flow controller (MFC) and a Harvard syringe pump respectively. The slug flow in the channel is recorded using a camera and the videos are analysed. The gas slug of pure CO_2 is found to decrease in size along the length of the channel due to absorption of gas in the liquid. This is also captured with the model developed and the mass transfer characteristics are studied. The pressure drop across the channel is determined by sum of the pressure drops from the gas slugs and the liquid plugs. A dimensionless correlation for the mass transfer coefficient is developed in terms of Sherwood number and compared with the existing correlations in the literature. They are found to be in close agreement with each other. In this case, due to the presence of chemical reaction, the enhancement of mass transfer is obtained. This is quantified with the help of an enhancement factor.

Keywordsâ€" Absorption, Enhancement factor, Mass transfer coefficient, Sherwood number

495024 Design and Evaluation of Thermodynamic Conditions for an Off-Shore Topside CO_2 Injection System

Abstract:

Carbon capture and storage (CCS) is one of the dominant technologies to tackle the global warming issue. The transport of CO_2 for geological storage is economically feasible by ship when the storage site location is off-shore and installment of an off-shore pipeline requires a huge capital cost. Ship transportation requires the captured CO_2 to be in liquid phase under pressurized thermodynamic conditions. The injection of liquid CO_2 into the geological reservoir involves pressurization and heating in order to maintain the safe well head operating conditions. Most of the recent studies focusing on ship transportation of CO_2 set their research boundary upstream of an injection system. On the other hand, studies investigating the storage mechanism in the EOR field or other geological locations assumed the fixed CO_2 conditions from the literature. The purpose of this study is to evaluate the design of CO_2 injection system for an off-shore storage site where CO_2 is available in the liquid phase at the vessel. Many previous studies considered the liquid CO_2 at the thermodynamic conditions of 6-7 bara and -52 °C assuming large CO_2 transport volumes. However, currently there are no commercial ships which transport liquid CO_2 at such cryogenic conditions and only small size ships up to 1,500 m^3 are being used for CO_2 transport at -20 °C and 20 bar. This study evaluates and compares the two thermodynamic conditions (7 bara and 20 bara) considering all the design and operational constraints of a liquid CO_2 injection system. This study also aims to fill the study gaps from the previous researches. For example, all the previous studies assumed the CO_2 stream as a pure liquid CO_2, however, practically this may not be possible since some amount of water is left in the CO_2 stream posing a threat of hydrate formation. In this study, we incorporate the CO_2 vapor return line to the pressure vessel which is an important aspect from safety point of view. This paper also investigates the application of two-stage rankine cycle to utilize the cold energy available from liquid CO_2 which can help to improve the overall efficiency of the injection system compared to that of the base case design. The working fluid employed in both the stages of rankine cycle is ammonia because of its highest power generation performance, easy availability and environment friendly characteristics compared to other working fluids. A well head injection pressure and temperature of 65 bar and 10 °C is set as a base case value for a storage depth of 1000 m to avoid any hydrate formation in the injection well or at the well outlet. The results show that the net specific power consumption for the two conditions (7 bara and 20 bara) is 4.52 and 1.90 kWh per ton of CO_2 injected respectively. Economic comparison has been performed to investigate the impact of operating conditions on the CO_2 injection process. The results show that the CAPEX for both the conditions is almost same since equipment involved in both the processes is the same. However, the OPEX for 7 bara design condition is approximately 2.4 times high compared to that of the 20 bara conditions. The high OPEX is attributed from the high operational energy requirement of the process. For a project life of 10 years, the specific cost per unit ton of CO_2 injected for the two design conditions (7 bara and 20 bara) came out to be 1.20 $ and 1.03 $ respectively. Finally, a sensitivity analysis has been done in order to investigate the effect of some important variables in the study.

503177 Keynote - Two-step Solarthermal Water Splitting – The Path Forward

Abstract:

Using concentrated sunlight to drive water splitting for renewable hydrogen production is the ultimate goal for a truly green energy world. The seminal paper, "Hydrogen and Oxygen from Water," *Science*, 197, 1050 (1977), by E.A. Fletcher and R.L. Moen ignited a flurry of activity in the field of solar thermalwater splitting. Focused research since then has resulted in substantial progress identifying particular chemical routes, active materials and various solarthermal reactor designs. This presentation will review key findings and will focus on the path forward emphasizing active materials and processes and identifying key areas of interest for future research.

491083 Syngas Production Via Solar Chemical-Looping Reformation of Methane in a Fixed-Bed Reactor

Abstract:

The production of syngas using concentrated sunlight via solar chemical-looping reforming (SCLR) is an approach to produce sustainable transportation fuels. SCLR is a two-step thermochemical cycle that produces separate streams of syngas and high-purity H_2. In the endothermic partial oxidation step, methane reacts with oxygen released from a metal oxide to produce syngas. In the exothermic water-splitting step, steam reacts with the reduced metal oxide to produce H_2. Solar energy is stored in the products, upgrading the calorific content of the feedstock by up to 28%. Combustion of the products emits 41% less carbon per kilojoule than combustion of the productions from non-solar CLR. The cycle is thermodynamically favorable at 1273 K and 1000 suns, conditions easily achievable by commercial solar concentrated optics used currently for electricity production. Based on a thermodynamic analysis of the process, the solar-to-chemical efficiency for SCLR, defined as the net increase in the higher heating value of the gases divided by the solar thermal input to provide process heat and parasitic work for gas pumping and separation, can reach 54% with complete conversion of CH_4 to syngas and H_2O to H_2.

While the net products and chemical energy requirements match that of solar steam reforming, decoupling the partial oxidation and H_2O-splitting steps provides numerous advantages. The products of SCLR are separate streams of 2:1 H_2:CO and high-purity H_2 compared to the single stream of 3:1 H_2:CO obtained from steam reforming. The 2:1 H_2:CO syngas matches the desired composition for the synthesis of liquid hydrocarbon fuels and H_2 is valuable as a fuel for electricity production via fuel cells or as a chemical feedstock in the synthesis of various chemical commodities such as ammonia. Furthermore, conventional steam reformers must operate with excess oxidizer (e.g. $H_2O/CH_4 = 3$) to prevent catalyst deactivation by carbon deposition. With solar CLR, carbon deposition can be prevented by limiting the reduction extent of the metal oxide.

In this paper, we present measured performance for SCLR in a prototype reactor that utilizes cerium dioxide (ceria) as the oxygen carrier material. The reactor is operated in the University of Minnesota high flux solar simulator. The reactor comprises two sections: a solar receiver/reactor and a heat recuperator. Concentrated sunlight enters the cylindrical receiver/reactor cavity section through an open circular aperture and heats a sealed assembly of concentric alumina tubes, referred to as the reactive element. The annulus of the reactive element is filled with 336 g of ceria particles with a bed void fraction of 45%. The particles are 5 mm (length and diameter), 78% porous and have a specific surface area of 0.114 m^2 g^{-1}. The reactive element extends beyond the solar cavity to the heat recuperator. In the heat recuperator, the annulus and inner tube of the reactive element are filled with alumina reticulated porous ceramic that is 85-90% porous and has pore densities of 5 and 10 pores per inch in the inner tube and annulus, respectively.

The cycle is accomplished by alternating the gas feed to the reactive element between CH^4 and H^2O. During the endothermic step, CH_4 flows through the inner tube, reverses direction at the closed end, and flows over the ceria particles in the annulus. Sensible heat from the product gases is transferred to the inlet flow in the heat recuperator. At the end of the partial oxidation step, the gas flow is switched automatically to H_2O. The operating conditions (temperature, reactant feed rates, reaction times) are selected based on a thermodynamic process model of the reactor. The model predicts CH_4 and H_2O conversions and H_2 and CO selectivities data from prior benchtop experiments.

Measured results include radiative input, transient temperatures of the solar cavity, reactive element, and inlet and outlet gases, upstream and differential pressures across the reactive element, and product gas composition. The axial temperature distribution along the top surface of the reactive element in the receiver/reactor and surface temperature of the reactor cavity are measured by alumina-sheathed type-B (Pt-30% Rh/Pt-6% Rh) thermocouple probes. The axial temperature distribution along the reactive element in the heat recuperator section and the gas temperatures at the inlet and outlet of the active reactive element are measured with type-K (Ni-Cr/Ni-Al) thermocouples. The high-flux solar simulator is controlled to obtain a cycle-averaged reactive element temperature of 1273 K. The radiative solar input is measured using a water-cooled black body calorimeter. The product gas composition is measured at 1 s intervals using Ramen laser gas spectroscopy. Data are evaluated to determine the solar-to-chemical efficiency of the reactor. The reactor demonstration provides an important proof of concept for SCLR and a comparison of measured efficiency to prior theoretical predictions of efficiency.

486955 Solar Gasification in a Molten Salt Reactor for Continuous Production of Syngas

Abstract:

The design and operation of a concentrated solar reactor to gasify biomass in a ternary eutectic blend of alkali metal carbonate salts are presented. Solar gasification has numerous advantages over conventional gasification. The use of concentrated solar energy as the source of process heat eliminates the need for partial combustion of the feedstock. The product synthesis gas is not diluted by excess CO_2 and N_2 nor contaminated by other combustion by-products. There is no need for an economically and energetically expensive oxygen plant that otherwise is necessary for oxygen-based partial combustion. The energetic value of the product is greater than that of the feedstockâ€"the difference is equal to the storage of solar energy in the fuel. The benefits of the molten salt solar gasifier in comparison to other solar gasifiers are the thermal mass of the salt allows continuous operation during solar transients, provides rapid heat transfer, catalyzes the gasification reaction, and reduces the amount of tars and other contaminants in the product gas stream for an operating temperature of 1200 K. In prior work, utilizing cellulose as feedstock and carbon dioxide as oxidizer, the reactor achieved a solar efficiency of 30% and converted 47% of the carbon in a continuous process. The efficiency at 1218 K was comparable to the highest achieved in other solar gasifiers operated at higher temperature, but the pneumatic feed system was prone to blockage and required a stoichiometric oxidizer ratio of eleven. The use of excess carbon dioxide increased the sensible heating requirement and exacerbated the entrainment of fine char particles in the product gas stream prior to complete conversion.

In the present study, we present a model of an improved volumetric pumping feed system and provide operational data for gasification of cellulose in the improved reactor. The feed system consists of a single fluted screw within a straight barrel, similar to that used for polymer extrusion systems. The feed enters the reactor as a densified cylinder rather than as loose powder, encouraging the formation of larger char particles which have a reduced tendency to be entrained by the product gases leaving the reactor. The design of the feed delivery system is based in part on a numerical model of the transient heat transfer and kinetics of pyrolysis and gasification of the feedstock material during transport.

The reactor was operated for CO_2 gasification of cellulose in the University of Minnesotaâ€™s high flux solar simulator. The simulator consists of seven radiation units, each composed of a 6.5 kW xenon short arc lamp close-coupled to a custom reflector in the shape of a truncated ellipsoid of revolution. The total power delivered to the aperture of the reactor was measured using a water-cooled black body calorimeter of the same aperture diameter. The feedstock is ash-free microcrystalline cellulose ($C_6H_{10}O_5$) sieved to a particle size of 0.5 mm. Temperatures of the solar cavity and within the salt melt are measured using Type-K thermocouples. The product gas composition is measured using a Raman Laser Gas Analyzer. The downstream gas lines and HEPA filter are analyzed for residual secondary products. Tars are separated from solids using a solvent wash and the particle size distribution of the solid material is measured using laser diffraction.

The reactor performance is quantified by the solar to fuel efficiency defined in rate form as the ratio of the lower heating value of the useful products and the sum of the solar input and the LHV of the carbonaceous feedstock. Carbon conversion is given by the ratio of the net gaseous carbon released from the reactor to the feedstock carbon delivered to the reactor. The measured species flow rates are compared to the equilibrium flow rates predicted using Gibbs free energy minimization. The data demonstrate the improvements in performance for the reactor with the new feed system in terms of longer operation without blockage, and higher solar efficiency.

494979 An Evaluation of Feedback and Model Predictive Control for the Rejection of Weather Transients in Renewable High-Temperature Solar Thermal Chemical Processing on a Novel Hybrid Solar-Electric Reactor

Abstract:

Concentrated solar power, which has been commercialized for electrical generation, is now of interest as a means of directly driving chemical processes. Investigations have included the use of solar heat to produce metals from oxide ores, generate renewable hydrogen for use as fuel, and to gasify biomass. These processes often operate at high temperatures where reactors and receivers are built partly or wholly from ceramics, ceramics that are vulnerable to thermal shock and fatigue. Weather transients, which can deprive a facility of solar heat within seconds, must be controlled to insure the longevity of solar-driven chemical process equipment. Supplementary electrical and/or gas heaters can be used to provide auxiliary heat when transient cloud cover would otherwise interrupt solar thermal chemical production. We evaluate different control schemes on a new solar-electric reactor prototype at the University of Colorado Boulder. These include both feedback and model predictive control paradigms, whose performance was measured by integral squared absolute error over the course of programmed weather transients on a 45kWe high-flux solar simulator. Implications for the implementation of commercial solar thermal processes will be discussed.

494571 Solar Hybrid Photo-Thermochemical Sulfur-Ammonia Water-Splitting Cycle: Photocatalyst, Thermodynamics and Plant Analysis

Abstract:

The solar driven hybrid sulfur-ammonia water splitting cycle (HySA) is a promising technology for the environmentally sustainable production of hydrogen fuel. It is a state-of-the-art process that integrates a solar-photocatalytic hydrogen production step (driven by the photonic portion of solar irradiance) with a high-temperature solar thermochemical oxygen evolution step (driven by the thermal portion) and efficient thermal energy storage as an integral part of the cycle. This work summarizes the recent findings on the two subcycles and the overall process design and optimization, following detailed experimental and numerical calculations. First, multiple photocatalysts, for the hydrogen production step, were assessed and characterized, finally, a binary band gap photocatalyst based on ZnS-CdS core shell particles doped with Pt was selected. Second, the materials and their properties for the oxygen evolution step were verified using thermogravimetric experiments, that supported a detailed thermodynamic analysis for the selection and design of the overall process. Finally, all information was assembled in one process that takes advantage of the hybrid solar photo-thermo character of the cycle and the incorporation of a portion of the process-fluids to design an integrated thermal-storage cycle. The solar to H_2 efficiency of the process was estimated up to 25% depending on reactor configurations.

495376 On the Potential for Enhancement of Solar Thermochemical Synthesis Processes By Nonequilibrium Plasma

Abstract:

The sustainable synthesis of fuels and chemicals from abundant low-value feedstock, such as carbon dioxide, methane, and water, is a key technological and global challenge. Solar thermochemical synthesis processes directly utilize the most abundant form of renewable energy, but face severe challenges due to the low reactivity of most feedstock gases to solar radiation and the intermittency of the solar resource. These challenges often lead to limited process efficiency and resiliency, which translate into systems with higher capital and operating costs. The combination of a solar thermochemical process with a non-solar co-process can be an effective option to achieve continuous operation, improve resiliency, and reduce costs. Moreover, if the concurrent operation of the solar and non-solar processes leads to greater overall efficiency, then the combined approach could have greater viability than either process alone. This talk discusses the potential of the use of nonequilibrium plasma as a co-process to solar thermochemical synthesis. The combined solar-plasma approach, denoted as Plasma-Enhanced Solar Energy (PESE), has the potential to enhance the efficiency and resiliency, and consequently improve the viability, of solar thermochemical synthesis processes.

Nonequilibrium plasmas are partially ionized gases in which the temperature of the free electrons is significantly higher than that of the heavy-species (molecules, atoms, ions). The electron temperature in atmospheric-pressure nonequilibrium plasmas are typically on the order of 1 to 2 eV, commensurate with the threshold energies in molecular bonds, whereas the temperature of the heavy-species range between room temperature to a few thousand degrees. Therefore, electrons and heavy-species are not in state of Local Thermodynamic Equilibrium (LTE), i.e., they are in a non-LTE (NLTE) state. Plasma-based chemical synthesis processes use electricity to directly modify gas chemistry and produce activated species (radicals, excited states, ions). These processes can achieve high energy efficiencies by selectively transferring energy to different channels. For example, microwave plasmas have shown over 90% energy efficiency (i.e. electrical to chemical) for the decomposition of carbon dioxide thanks to effective step-wise vibrational excitation, the most effective channel for dissociation [1]. Moreover, the reliance of plasma processes on electrical energy makes them ideal for continuous operation, but also limits their economic viability and their net sustainability benefit, even if the required electricity is obtained from renewable energy sources, such as wind and solar.

PESE chemical synthesis processes seek to combine the advantages of solar thermochemical and plasma processes. The potential advantage of PESE to improve the efficiency of solar thermochemical synthesis is due to the fact that the molecular excitation produced by free electrons in non-equilibrium plasmas can significantly increase solar photon absorption leading to the enhancement of chemical reaction kinetics. Moreover, PESE chemical synthesis processes could rely on electrical energy to compensate for fluctuations in the solar energy input (during daytime) or even substitute it (at nighttime), and therefore ensure continuous operation.

Preliminary computational analyses of solar radiation absorption through a carbon-dioxide (CO_2) mixture in a state of NLTE reveal over two orders of magnitude increase in absorption compared to the same mixture in a state of LTE, achieving 50 to 80% radiative energy absorption efficiency. The calculations were performed with the line-by-line nonequilibrium radiation code SPARTAN [2, 3] using 32 types of atomic and molecular transitions. The results show that CO_2 in LTE absorbs negligible solar irradiance (mostly UV and far IR), whereas CO_2 in NLTE displays significantly more absorption, especially in the visible range, but also emission, which diminishes the overall amount of absorbed radiation. Both findings are consistent with established understanding of radiative transport; however, the novelty is finding that the net effect is a significant increase in solar energy absorption. The results also show that CO_2 tends to behave as a perfect absorber of solar energy for very high degrees of NLTE (i.e. larger differences between electron and heavy-species temperatures). This fact, although not intuitive given that plasmas are known to be good emitters (e.g., radiation from arc lamps), can be understood by considering that a medium is prone to absorb the same type of radiation it emits. Furthermore, it is to be noted that the concept of a blackbody, a perfect absorber and emitter or radiation, corresponds to a medium in LTE. In fact, the approximation of the solar spectrum as the emission from a blackbody at ~ 5780 K is a consequence that the sun's photosphere is largely regarded as being in LTE. Therefore, a NLTE medium provides the possibility to control its radiative properties by tuning its degree of thermodynamic nonequilibrium.

To evaluate the potential of PESE, a solar receiver-reactor system equipped with two types of nonequilibrium plasmas, namely a gliding arc discharge and a microwave plasma, is being built at UMass Lowell. The PESE chemical synthesis reactor system is designed to evaluate comparable solar and electrical energy inputs at the 1 kW level, spanning the whole range of combined solar-plasma operation, i.e. from 100% solar to 100% electrical energy inputs. The microwave plasma has the potential for the greatest energy efficiency, whereas the gliding arc discharge has simpler operation, more stability, and greater potential for scalability. Design challenges being addressed include the establishment of a stable plasma volume within the focal point of the concentrated solar radiation from a high-flux solar simulator [4], as well as the maximization of the plasma volume and the gas-plasma interaction region, the incorporation of a downstream catalytic monolith, among others.

Acknowledgements

The authors gratefully acknowledge support from the U.S. National Science Foundation through award CBET-1552037.

References

[1] A. Fridman, *Plasma chemistry*, Cambridge University Press, New York (2008)

[2] M. Lino da Silva, *SPARTAN - Simulation of PlasmA Radiation in ThermdynAmic Nonequilibrium*, http://esther.ist.utl.pt/spartan/

[3] M. Lino da Silva, *An adaptive line-by-line—statistical model for fast and accurate spectral simulations in low-pressure plasmas*, J. Quant. Spectrosc. and Rad. Transfer 108 (2007) 106–125

[4] S. Bhatta, D. Nagassou, J. P. Trelles, *Solar photo-thermochemical reactor design for carbon dioxide processing*, Solar Energy 142 (2017) 253–266

496114 Direct Two Step Solar Metallothermic and Electrolytic Production of Rare Earth Elements from Oxides

Abstract:

Rare Earth Elements (REEs) exhibit a range of unique electronic, optical and magnetic properties which make them attractive for many niche applications, including metal alloys, catalysts, petroleum refining, glass additives, polishing compounds, and ceramics. The demand for rare earth elements is expected to grow in the coming years, especially in uses related to critical clean energy technologies such as PV devices, batteries, high-performance magnets, and phosphors for lighting and electronics. Cerium (Ce), lanthanum (La), neodymium (Nd), and praseodymium (Pd) account for ~86% of the total global REE consumption. The current electrolytic production route through molten chlorides is quite inefficient due to the stability of Re^{+2} ion in chloride melt, whereas the oxide-fluoride route suffers from low solubility of oxides in fluorides. Additionally, these extractive processes are currently driven by fossil fuel based energy sources with very high greenhouse gas impacts and resource depletion potential.

Energy consumption and environmental concerns, coupled with the fact that the current electrolytic production method must occur at high temperatures, provides significant motivation for investigating solar driven metallothermic/ electrolytic processing routes as alternatives to conventional REE production for providing â€œcleanâ€ high temperature process heat and electricity. In this work, a new two-step REE production route consisting of direct metallothermic reduction of RE-oxides using Ca and subsequent electrolytic recovery of metallic Ca in a mixed chloride bath has been explored from the theoretical energy consumption perspective. The new process is critically evaluated using thermodynamic analysis and compared with the state-of-the-art. The two-step process shows promise in terms of significant improvement in the energy efficiency compared to the current electrolysis routes. The potential technological challenges in scale-up and impact of solar variability will be discussed as well.

494169 Analytical Analysis of Flow Distribution Uniformity As a Design Criteria for a Novel Micro Packed-Bed Bionic Reactor

Abstract:

The Micro Packed-bed Bionic Reactor (MPBR) is a novel configuration aiming for fast catalytic gas-solid reactions. Hierarchical distributed fluid channel and microscale permeable framework packed with small particle catalysts can realize a higher reaction effectiveness, a lower pressure drop and a better heat transfer.The flow distribution uniformity substantially affect the performance of MPBR, which is determined by operating conditions and especially size parameters. This work endeavors to use analytical methods to seek a theoretical instruction of structure design.

The elementary unit of MPBR consists of a catalyst bed with two halves of fluid channel on both sides, which is governed by Bermanâ€™s equation in fluid channels and Darcyâ€™ law in catalyst bed. The one-dimensional simplified mathematical model is obtained by integration of channel equations along radial direction and coupling through Darcyâ€™s flow equation. The resulting expression turns out to be a second order homogeneous linear differential equation of pressure difference between the two fluid channels, and the analytic solution is the function of three dimensionless parameters, the inlet Reynolds number, the aspect ratio of fluid channel and the unit permeability. The analytic solution makes it possible to quantitatively estimate the distribution uniformity by the relative standard deviation of pressure difference, which is proved, under reasonable conditions, to be linear to P_{re} , a new dimensionless parameter extracted from the analytic solution indicating the relative proportion between the pressure drop of fluid channel and catalyst bed. Deduction demonstrates that P_{re} equals the product of the three aforementioned dimensionless parameters, meaning that giving a maximum restriction for the relative standard deviation, the operating conditions and structure size can be designed correspondingly.

Further numerical simulations have been carried out to estimate the accuracy of the analytic solution, and verify that the error is tolerant even in certain parameter ranges where some of the assumptions the analytic solution is based on is invalid.

496279 Macroscopic and Microscopic Modeling of a Lithium-Ion Cell with Solid-State Intercalation Effect

Abstract:

The macroscopic and microscopic modeling of lithium-ion cells considers electrochemical transport, kinetics of electrode reactions and porous electrode kinetics, and solid state diffusion. The intercalation/deintercalation process of lithium at solid particle surface is the key to understand the charge-discharge phenomenon of lithium battery. The solid state diffusion model with a two-component diffusion term can be reduced to a simple nonlinear diffusion form and applied to the behavior of single electrode particle. The nonlinear diffusion equation shows the vacancy effect in the solid-state diffusion of the particles inside the cell is illustrated in the nonlinear diffusion equation. The simulated results of voltage versus capacity obtained at discharge rate below 0.6C of C-rate agree well with the experimental data. Multiple reactions may also be included in the model.

489446 Radiation Induced Grafted Fep-g-Polystyrene Membranes: Chemical Degradation and Quantum Chemical Simulation

Abstract:

Polymer electrolyte membrane fuel cells (PEMFCs) have received significant attention in recent years due to high efficiency and low emissions. The proton exchange membrane (PEM) is one of the key components that of significant importance to the fuel cell performance, which is electron insulating while proton conductive and should prevent the crossover of fuels and oxidant, as well as stable in the fuel cell operation environment[1]. Radiation induced graft (RIG) is a versatile method to modify pre-existing polymers to introduce a variety of desired functionalities, and has been studied to prepare PEM[2].

In this work, PEM was prepared by RIG polymerization of styrene onto FEP base polymer followed by sulfonation using chlorosulfonic acid. The graft polymerization was performed by immersing the FEP films into a series of styrene solutions and submit to the ^{60}Co radiation source. The sulfonated FEP-g-polystyrene membranes exhibited much higher proton conductivity than Nafion 211 as shown in Fig.1(a). And the membrane prepared in 40%v/v solution showed better fuel cell performance than Nafion 211 under the same test conditions in Fig.1(b).

Fig.1 (a) Proton conductivity of the sulfonated FEP-g-polystyrene grafting in solutions of different concentration in comparing with Nafion 211. The corresponding degrees of graft were listed alongside. (b) Polarization curve of the sulfonated FEP-g-polystyrene in comparison with Nafion 211, catalyst loading is $0.6mg_{Pt}/cm^2$ and H_2/air was used in the fuel cell test.

Also elucidation of the mechanism of membrane degradation is an active area of research. Current papers have proposed a degradation mechanism that involves attack of oxygen radicals on susceptible groups in the polymer backbone [3].In this work, Fenton¡¯s test was conducted using 30% H_2O_2 and 10ppm Fe^{2+}. And the membranes were also tested in fuel cell operation conditions. The resulting solution of Fenton test and water collected at the anode and cathode were analyzed for fluoride ion content using an ion chromatography system. XPS, FTIR and SEM were also used to characterize the degraded membrane.

Previous work has suggested that hydroxyl free radicals can attack polymer end groups having H-containing terminal bonds (such as $-CF_2COOH$) that are formed during membrane processing[4]. Here we would use quantum chemical simulations to elucidate degradation mechanism of the sulfonated FEP-g-polystyrene membrane suffered to the ¡¤OH and ¡¤OOH radical attack by searching the transition state of different reaction pathways.

Keywords: radiation induced graft, FEP-g-polystyrene, chemical degradation, quantum chemical calculation

Acknowledgements: This work was supported by the Intergovernmental International Scientific and Technological Innovation Cooperation Key Projects (2016YFE0102700) and National Natural Science Foundation of China (51573083).

References:

[1] Zhang L, Chae S, Hendren Z, Park J, Wiesner MR. Recent advances in proton exchange membranes for fuel cell applications. Chem Eng J, 2012, 204-206: 87-97.

[2] Gubler L. Polymer Design Strategies for Radiation-Grafted Fuel Cell Membranes. Advanced Energy Materials, 2014, 4(3): 1441-1443.

[3] Healy J, Hayden C, Xie T, Olson K, Waldo R, Brundage A, Gasteiger HA, Abbott J. Aspects of the Chemical Degradation of PFSA Ionomers used in PEM Fuel Cells[J]. Fuel Cells, 2005, 5(2):302-308.

[4] Curtin DE, Lousenberg RD, Henry TJ, Tangeman PC, Tisack ME. Advanced materials for improved pemfc performance and life. Journal of Power Sources, 2004, 131(1-2), 41-48.

486864 Analyses on the Effect of Natural Gas's Supercooling Temperature on the Liquefaction Performance of Prico Process

Abstract:

Due to the advantages of cleanness, high calorific value and so on, LNG (liquefied natural gas) is becoming increasingly important in the energy structure with the changes of energy utilization situation and the development of technology. However, natural gas liquefaction processes are complicated with significant interactions of each component, making it difficult to obtain the optimal working condition. Meanwhile, in the mixed refrigerant liquefaction processes, compositions of the mixed refrigerant and other working parameters have predominant effects on the performance of compressor and the matching of hot/cold composite curves, which may result in large exergy losses of some components and low refrigeration efficiency if deviated from the optimum. Especially the natural gas subcooling degree varies in different literature, the optimization and determination of which needs further investigations.

Based on the simplified prico process, this paper focuses on the effects of the initial pressure and subcooling degree of the natural gas (the cooling temperature of liquefier) on the liquefaction performance such as the specific/unit liquefaction work and the specific/unit volumetric liquefaction capacity. The main contents are presented as follows.

(1)The mixed refrigerant composed of R600a, R290, R1150, R50 and R728 is applied.Pure methane is used to simulate natural gas. Aspen HYSYS was used to simulate the natural gas liquefaction process with an initial pressure range of 0.5 MPa ~ 5MPa.

(2) the effects of the initial pressure of natural gas and the cooling temperature of liquefier on the specific liquefaction work and specifc volumetric liquefaction capacity were analyzed . Results show that different initial pressure and cooling temperature of liquefier have significant effects on system performance.

(3) Through the comprehensive consideration of performance and cost optimization, the method for the determination of the natural gas subcooling degree of the prico process we proposed, which was of great significance to the subsequent experiment and application research.

501748 Understanding the Effect of ZINC and Achieving LONG CYCLE Life in Cu-Intercalated Bi-Birnessite/ZINC Batteries

Abstract:

UNDERSTANDING THE EFFECT OF ZINC AND ACHIEVING LONG CYCLE LIFE IN Cu-INTERCALATED Bi-BIRNESSITE/ZINC BATTERIES

Manganese dioxide (MnO_2)/Zinc (Zn) batteries have high volumetric energy densities > 400Wh/L (1), which make them ideal for use in grid applications as safe, low cost and non-flammable energy storage devices. However, their use has been curtailed to primary applications because of the irreversible nature of MnO_2 beyond 5-10% of its 2nd electron capacity of 617mAh/g (2). Recently, we solved the issue of MnO_2 rechargeability by using a layered polymorph called birnessite mixed with bismuth oxide intercalated with Cu ions (Cu intercalated Bi-birnessite) (3). The Cu-intercalated Bi-birnessite was shown to cycle near the 2nd electron capacity for thousands of cycles. For energy dense batteries where areal and volumetric capacities are important parameters, it was also shown to cycle at very high areal capacities of 10-29mAh/cm^2 against a sintered NiOOH counter-electrode for thousands of cycles. However, for true applicability in practical energy dense batteries its pairing with a Zn anode is essential. The use of Zn anodes has also presented problems as it is the source of zincate ions in electrolyte that react with the cathode, MnO_2, to form electro-inactive phase called haeterolite ($ZnMn_2O_4$) (4,5). The best reported cycle life data for high depth-of-discharge (DOD) birnessite cathodes with Zn anodes had been 50 cycles (4) till our recent publication, which showed over 90 cycles (3).

In this presentation, we report the effect of zincate ions on the Cu-intercalated Bi-birnessite cathodes beyond 100 cycles (6). The Cu-intercalated Bi-birnessite cathodes when paired with Zn anodes are shown to deliver 160Wh/L and cycle reversibly for over 100 cycles. The Cu ions play an important role in mitigating the detrimental effect of zincate ions in the 100 cycles; however, the zincate ions eventually poison the cathode to form $ZnMn_2O_4$. The mechanism through which $ZnMn_2O_4$ is formed is presented in detail with the aid of electroanalytical and spectroscopic methods. A solution of trapping the zincate ions is also presented, where the membrane that is used successfully traps the zincate ions from interacting with the cathode and thus, extend cycle life to over 900 cycles as shown in Figure 1. This is the best reported cycle life data with a manganese dioxide cathode accessing the near 2nd electron capacity paired with Zn anodes.

Figure 1. A Cu-intercalated Bi-birnessite cathode paired with Zn anode achieving cycle life greater than 900 cycles with specialized membrane.

Reference:

1] Gallaway, J. W.; Hertzberg, B. J.; Zhong, Z.; Croft, M.; Turney, D. E.; Yadav, G. G.; Steingart, D. A; Erdonmez; C. K.; Banerjee, S. *"Operando identification of the point of [Mn$_2$]O$_4$ spinel formation during γ-MnO2 discharge within batteries"* Journal of Power Sources **321,** 135-142 (2016).

2] Ingale, N. D.; Gallaway, J. W.; Nyce, M.; Couzis, A.; Banerjee, S., *"Rechargeability and economic aspects of alkaline zincmanganese dioxide cells for electrical storage and load leveling,"* Journal of Power Sources **276**, 718 (2015)

3] Yadav, G. G.; Gallaway, J. W.; Turney, D. E.; Nyce, M.; Huang, J.; Wei, X.; Banerjee, S. *"Regenerable Cu-intercalated MnO_2 layered cathode for highly cyclable energy dense batteries"* Nat. Commun. **8**, 14424 (2017).

4] Bai, L.; Qu, D. Y.; Conway, B. E.; Zhou, Y. H.; Chowdhury, G.; Adams, W. A. *"Rechargeability of a Chemically Modified MnO2/Zn Battery System at Practically Favorable Power Levels"* J. Electrochem. Soc. **140**, 884-889 (1993).

5] Huang, J.; Yadav, G. G.; Gallaway, J. W.; Nyce, M.; Banerjee, S. *"Calcium Hydroxide Membrane As a Separator to Immobilize Zincate Ions in Secondary Alkaline Batteries"* ECS Meeting, Abstract MA2016-01 490 (2016).

6] Yadav, G. G.; Wei, X; Huang, J.; Gallaway, J. W.; Turney, D. E.; Nyce, M.; Secor, J.; Banerjee, S., paper submitted (2017).

500366 Integration of Experimental and Modeling for Evaluation of Microalgae Culturing in Split Airlift Photobioreactor

Abstract:

Â Integration of Experimental and Modeling for Evaluation of Microalgae Culturing in Split Airlift Photobioreactor

Laith S. Sabri[1], Abbas J. Sultan[2] and Muthanna H. Al-Dahhan[3]

[1, 2, 3] Multiphase Reactors and Applications Laboratory mReal.

Department of Chemical and Biochemical Engineering, Missouri University of Science and Technology, Rolla, MO 65409-1230. USA

Abstract

The movement of microalgae cells inside the photobioreactors is the key to enhancing the biomass productivity. This movement could affect not only by the hydrodynamics (flow dynamics) inside the photobioreactor but also the light intensity which is received. In fact, when the microalgaeâ€™s cell received the light from the illumination surface and moving to the dark zone, biomass productivity could frequently become high. Moreover, the availability of the light in all the zones inside the reactor plays a critical factor for photobioreactor performance due to the effect of the biomass enrichment. Unfortunately, the interactions between the hydrodynamics parameters and the light availability remain unclear yet. Therefore, the request for both the mechanics of cell motion and for the mathematical model which contact between the variable concentrations of microalgae cells with irradiance spread inside the reactor is need. This work focuses on the integrated of the cells' movements (cell positions) which are tracked by using an advanced non-invasive technique called Radioactive Particle Tracking (RPT) with a mathematical model for growth rate. This model linked the irradiance distribution to the concentrations of cells through the period of growing. A Plexiglas split photobioreactor column carried out in this experimental work with 5.5-inch (0.14m) inner diameter. A green alga (Scenedesmus) used for the cultivation at different superficial gas velocities (1, 2, and 3 cm/sec). The advanced information which obtained from this investigation crystallizes the understanding and knowledge of photobioreactor behavior. The experimental results and conclusions will present at the conference.

Keywords: RPT, Split Airlift Photobioreactor, Microalgae, Growth rate model.

Thermodynamic analysis of an integrated Ca-Cu chemical loop for abatement of ventilation air methane

Yongxing Zhang[1,*], Xiaoling Chen[2], Behdad Moghtaderi[3]

[1]National Engineering Laboratory for Pipeline Safety/MOE Key Laboratory of Petroleum Engineering,
[2]Beijing Key Laboratory of Process Fluid Filtration and Separation, College of Mechanical and Transportation Engineering, China University of Petroleum-Beijing, China
[3]Priority Research Centre for Frontier Energy Technologies & Utilisation, Chemical Engineering, School of Engineering, Faculty of Engineering & Built Environment, The University of Newcastle, Australia
*Corresponding author: yongxing.zhang@uon.edu.au

Abstract

Methane is a major greenhouse gas (Global Warming Potential is 25 times that of CO_2) and fugitive methane (CH_4) from coal mines is the major source of anthropogenic CH_4. Approximately 64% of methane emissions in coal mine operations are in the form of ventilation air methane (<5 vol. % methane in air). An integrated Ca-Cu chemical loop is proposed to utilize ventilation air methane with inherent CO_2 capture as demonstrated in the below figure. Similar with a common chemical looping concept, it mainly consists of two reactors, an oxidation reactor (OR) and a calcination reactor (CR), between which the metal/metal oxides are circulated. In the oxidation reactor, copper is oxidised to its oxide form (mainly CuO), which is then acted as the catalyst for the catalytic combustion of ventilation air methane. Meanwhile, the product of CO_2 is adsorbed by the adsorbent of CaO and calcium carbonate ($CaCO_3$) as formed. The mixture of CuO and $CaCO_3$ is then transferred to a reduction reactor, in which CuO is reduced to Cu by high purity methane. Meanwhile, the endothermic calcination reaction of $CaCO_3$ occurs and the thermal heat is provided by the exothermic reaction of CuO reduction. CO_2 and steam are the main products from CR and the product gas from OR primarily consists of N_2, steam and excess oxygen. As noted, CO_2 and N_2 do not mix in the proposed process. Most importantly, both OR and RR are in autothermic operation and as such no extra energy is consumed for the strong endothermic calcination process, causing a large amount of efficiency reduction otherwise.

A detailed thermodynamic analysis of the process was carried out to evaluate its technical viability and overall performance. Taking into account the CO_2 capture efficiency, electricity efficiency, methane consumption and reduction ratio of GWP together, the most suitable operation parameters for the process were determined. Specifically, the temperature for CR and RR were set to be 625°C and 850°C respectively. It was also found that the process was able to work with various concentrations of ventilation air methane, but it should be noted that the CO_2 capture efficiency was less than 50% for the methane concentration lower than 1 vol. %. For ventilation air methane with methane concentration in the range of 3 to 4 vol%, high CO_2 capture efficiency, electricity efficiency and reduction ratio of GWP could be obtained, especially a high purity of N_2 stream.

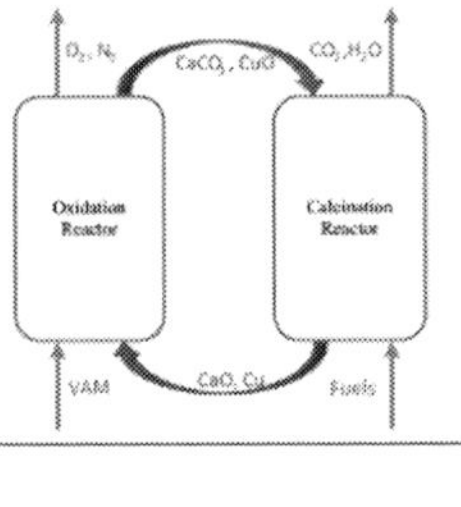

Objectives

Chemical Looping Combustion (CLC) is considered as one of the most promising technologies in capturing CO_2 from the processes of utilizing fossil fuel. Ventilation air methane, as a source of fuel, is hard to be used with inherent CO_2 capture in a traditional CLC process. The current study proposed a calcium-copper based CLC process, making it possible to efficiently utilize the ventilation air methane while capture CO_2 inherently, especially in an autothermic condition. The main goal of this project is to evaluate the performance of the designated process from a thermodynamics point of view and determine the optimal operation condition.

Advantages

- High Purity of CO_2 can be obtained when H_2O is condensed, thus eliminating the energy penalty for CO_2 separation.
- A small amount of pure CH_4 is needed to achieve a high CO_2 capture efficiency (higher than 90%) in an autothermic condition.
- The GWP (Global Warming Potential) index of ventilation air methane is reduced by a large margin.
- The process produce a high purity of N_2 (purity > 99, dry basis) as the oxygen is consumed by the oxidation of copper.

Results

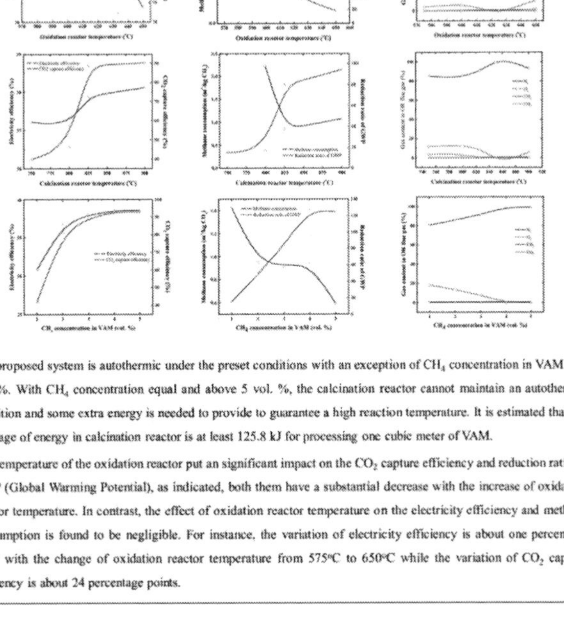

- The proposed system is autothermic under the preset conditions with an exception of CH_4 concentration in VAM at 5 vol. %. With CH_4 concentration equal and above 5 vol. %, the calcination reactor cannot maintain an autothermic condition and some extra energy is needed to provide to guarantee a high reaction temperature. It is estimated that the shortage of energy in calcination reactor is at least 125.8 kJ for processing one cubic meter of VAM.
- The temperature of the oxidation reactor put an significant impact on the CO_2 capture efficiency and reduction ratio of GWP (Global Warming Potential), as indicated, both them have a substantial decrease with the increase of oxidation reactor temperature. In contrast, the effect of oxidation reactor temperature on the electricity efficiency and methane consumption is found to be negligible. For instance, the variation of electricity efficiency is about one percentage point with the change of oxidation reactor temperature from 575°C to 650°C while the variation of CO_2 capture efficiency is about 24 percentage points.
- Over the examined calcination reactor temperature range, the electricity efficiency, CO_2 capture efficiency and reduction ratio of GWP are increasing with the increase of temperatures. It is noticeable that the variations become slight for the temperatures above 825°C. For the plot of methane consumption versus calcination reactor temperature, it drops rapidly first from 2.21 m^3 at 800°C to 0.85 m^3 at 825°C for reducing one kilogram CO_2 emissions and then increase slightly to 1.07 m^3 at 900°C.
- The electricity efficiency, CO_2 capture efficiency and reduction ratio of GWP are increasing with the increase of methane concentration in VAM when it is lower than 4 vol. %. With a higher methane concentration, all of them begin to level off at the summit point. As to the methane consumption, it reduces with the increase of methane concentration in VAM as a whole. It should be noticed that the methane consumption is almost steady at 0.94 m^3 for the methane concentration in VAM varying from 2 to 4 vol. %.
- The purity of N_2 product is also related to the oxidation reactor temperature, calcination reactor temperature and CH_4 concentration in VAM. The most influential factor is methane concentration in VAM which leads to a reduction of 20 percentage points over the examined range and the follower is the factor of calcination reactor temperature.

Conclusions

- The thermodynamics analysis confirms that the proposed system is autothermic with some suitable parameters while high CO_2 capture efficiency, reduction ratio of GWP and low methane consumption can be achieved..
- 99% purity of N_2 can be obtained with the oxidation reactor temperature at 625°C, calcination reactor temperature at 850°C and CH4 concentration in VAM at 4 vol. %, which trades off the economy of the system.

References

[1] K. Shah, B. Moghtaderi, E. Doroodchi et al., A feasibility study on a novel stone dust looping process for abatement of ventilation air methane. Fuel Processing Technology, 2015, 140, 285-296.

[2] H. Lu, E. P. Reddy and P. G. Smirniotis, Calcium oxide based sorbents for capture of carbon dioxide at high temperatures. Ind. Eng. Chem. Res., 2006, 45, 3944-3949.

Acknowledgement and Contact

The authors acknowledge financial support from the Science Foundation of China University of Petroleum, Beijing (No. 2462014YJRC050).

493294 Measurements of the Size Distribution and Concentration of Engineered Nanoparticles in Aqueous Media: Methods and Applications

Abstract:

Engineered nanoparticles (ENP) are increasingly used in manufacturing processes and as additives to consumer products to enhance product properties. Understanding the environmental fate and impact of ENPs relies on the ability to characterize them at relatively low levels in complex media, which often contains additional naturally occurring nanoparticles. We present the application of single-particle inductively coupled plasma mass spectrometry (sp-ICP-MS) and a recently developed hyphenated technique combining electrospray aerosol generation, particle size selection using differential mobility analysis (DMA), and elemental analysis using inductively coupled plasma mass spectrometry (ICP-MS) to characterize gold and silver nanoparticles in aqueous media. These methods are capable of detecting nanoparticles of 20-100 nm diameter at part-per-billion (ppb) levels in environmentally relevant aqueous solutions. These methods were used in an inter-comparison of NIST standard reference materials (SRMs). Following instrument characterization, the DMA-ICP-MS and sp-ICP-MS was used to assess the fate of gold nanoparticles in a solution containing daphnia magna. Insights into dose verifications obtained by DMA-ICP-MS and sp-ICP-MS will be discussed, as well as their performance for the analysis of the NIST SRMs.

499140 CFD Simulation and Analysis of Gas-Liquid Segmented Flow with Mass Transfer in Microfluidic Devices: Case Study of CO_2 Gas Solubility and Diffusivity Measurement in Aqueous NaCl Solution

Abstract:

The microfluidic gas-liquid segmented flow has the potential for measurement of thermodynamic properties such as solubility, diffusivity, and reaction rates. In multiphase systems, mass transfer resistance across the phase boundaries controls the mass transfer rates and hinders determination of thermodynamic properties. Microfluidic gas-liquid segmented flow minimizes this problem by increasing the specific interfacial area and improving mass transfer efficiency across phase boundaries [1]. Moreover, microfluidic devices significantly reduce measurement equipment size, experimental resource requirement, and measurement time. Using both experiment and numerical simulation, we study the solubility and diffusivity of CO_2 in the aqueous NaCl solution. In the experimental work, we developed long serpentine channel microfluidic devices that may be used at various CO_2 pressures and liquid solution flow rates to achieve segmented flow conditions. CO_2 bubble size is measured (at 293K) along the length of the channel to determine CO_2 solubility and diffusivity. To accurately measure these properties, we require pressure profile along the length of the channel which can be accurately calculated through CFD simulations. In simulation work, we use CFD to simulate and analyze gas-liquid segmented flow in the same microfluidic designed device. CFD analysis allows precise examination of the pressure profile along the channel length, flow behavior and localized concentration field. These simulations use Volume of Fluid method to address gas-liquid phases paired with momentum balance and continuity equations to capture the flow behavior [2]. Mass transfer between vapor and liquid phases is addressed by the diffusion equation, and application of Henry's law at the phase boundaries is used to analyze temporal and spatial variations in CO_2 concentration. We show CFD simulation results for pressure profile and use it to calculate the CO_2 solubility and diffusivity. We also show that CFD can simulate and predict the solubility of CO_2 in different salt solutions and different geometry devices which will help in reducing the experimental effort.

References

[1] Günther, A. and Jensen, K.F. "Multiphase microfluidics: from flow characteristics to chemical and materials synthesis." *Lab on a Chip*, 6(12), 1487-1503, 2006.

[2] Hirt, C.W. and Nichols, B.D. "Volume of fluid (VOF) method for the dynamics of free boundaries." *Journal of Computational Physics*, 39(1), 201-225, 1981.

Growing Scientific Leaders With Concentrated Sunlight: A Tribute to E. A. Fletcher

Bob Palumbo
Prof. of Mechanical Eng.

Luke Venstrom
Asst. Prof. of Mechanical Eng.

VALPARAISO UNIVERSITY

Growing Scientific Leaders

By Developing Deep Science and Engineering Knowledge

By Developing Communication Skills

By Developing the Passion to Serve Others

VALPARAISO UNIVERSITY

Model for Engaging Undergraduate Students

Curricular	**Senior Design:**	• Two-semester course • Design during fall semester • Fabricate and test during spring semester • Multi-disciplinary teams with faculty advisor
	Credit:	• Research during academic year • Students earn up to 3 credits towards mechanical engineering degree
Extra-curricular	**Summer Research:**	• 10 to 12 weeks over the summer • Includes stipend and free housing in dorm • Intentional cross-disciplinary interaction

VALPARAISO UNIVERSITY

Undergraduate Student Activity

VALPARAISO UNIVERSITY

Valpo Solar Furnace

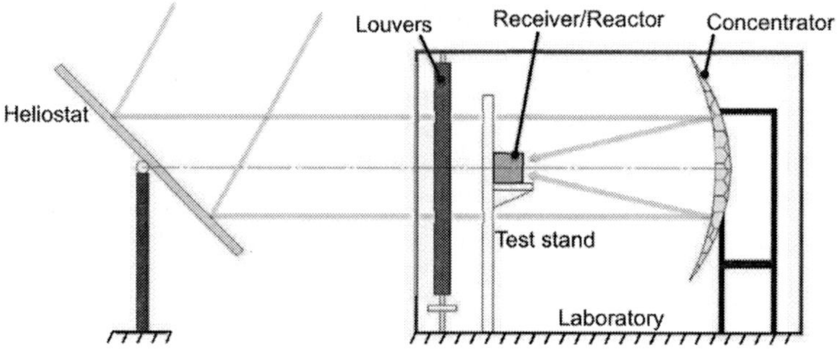

VALPARAISO UNIVERSITY

Valpo Solar Furnace

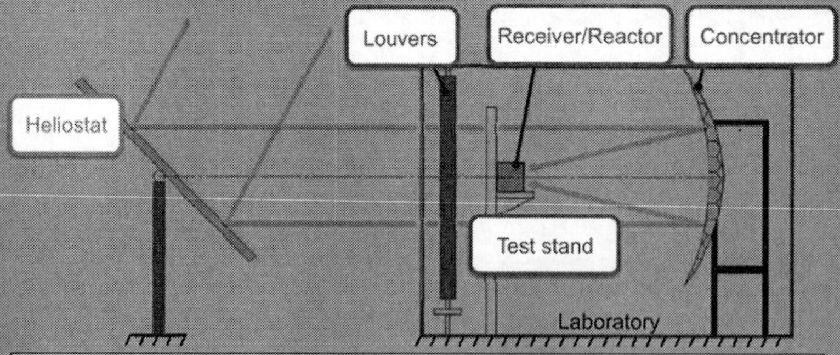

Designed and manufactured by undergraduate students

Undergraduate students contributed to design via analysis and construction of a prototype

VALPARAISO UNIVERSITY

Test Stand

Design Constraints:
- 3-axis motion accurate within .01"
- Support 250 lbf
- Integrate chilled water flow up to 10 gpm, pressurized air supply, and exhaust gas line
- Safety:
  - brake in event of power loss
  - emergency stop switches
  - laser safety scanners
- Able to hold numerous receivers:
  - Calorimeter
  - Lambertian target
  - Solar thermochemical reactors

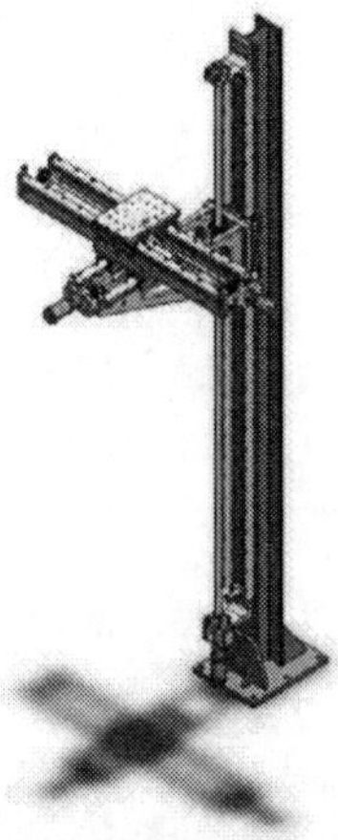

VALPARAISO UNIVERSITY

Test Stand

Analyzed by undergraduate students with FEA: Factor of safety of 11

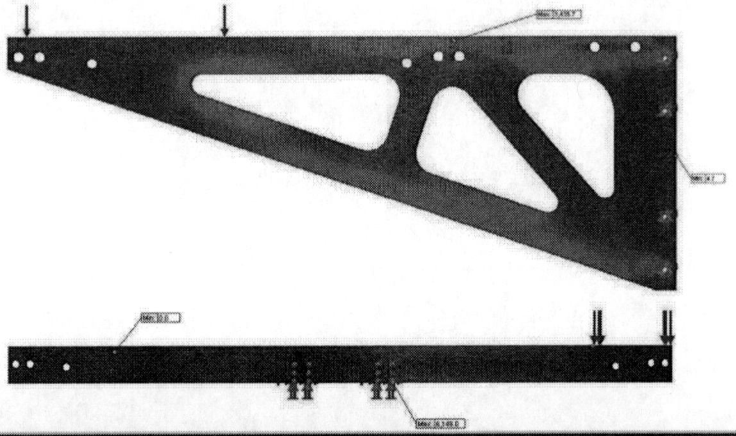

VALPARAISO UNIVERSITY

Test Stand

VALPARAISO
UNIVERSITY

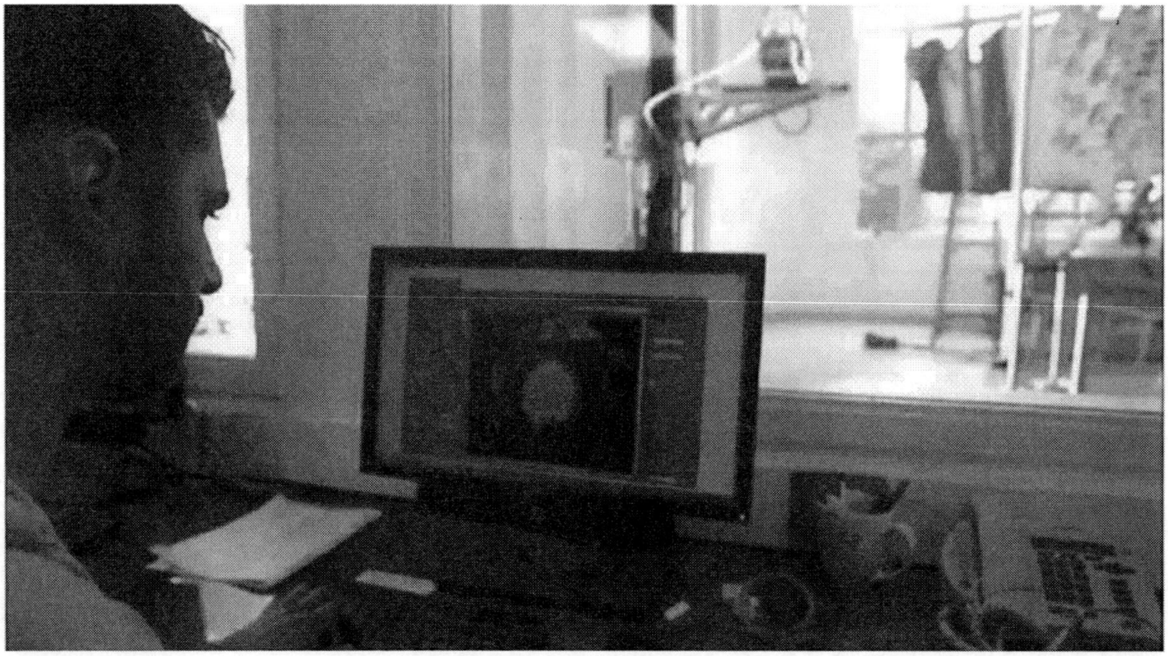

Solar Thermal Decoupled Electrolysis of Water

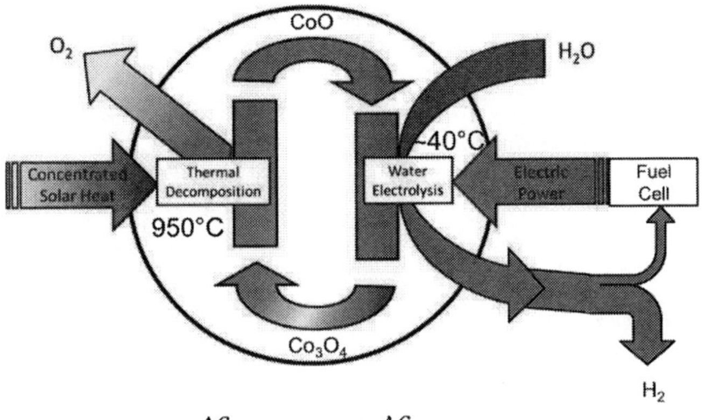

$$\eta_{cycle} = \eta_{abs}\frac{\Delta G_{H_2+0.5O_2=H_2O} - \Delta G_{Electrolysis}}{\Delta H_{Co_3O_4\rightarrow 3CoO+0.5O_2}} = 38\%$$

Palumbo *et al.*, 2012, *Chem. Eng. Sci.*, **84**, 372–380.

VALPARAISO UNIVERSITY

Rotary Drum Reactor Concept

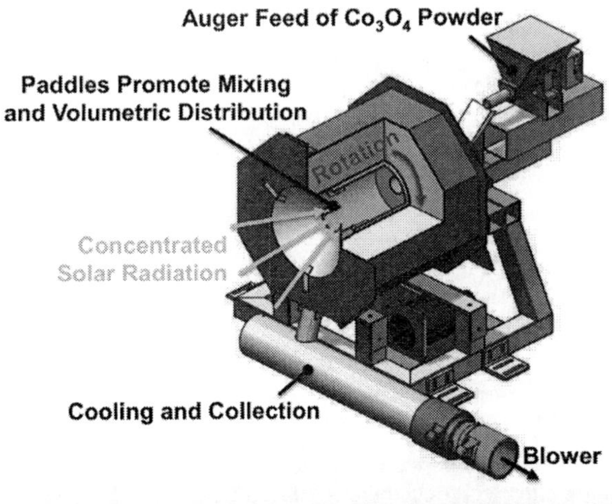

VALPARAISO UNIVERSITY

Rotary Drum Reactor Model

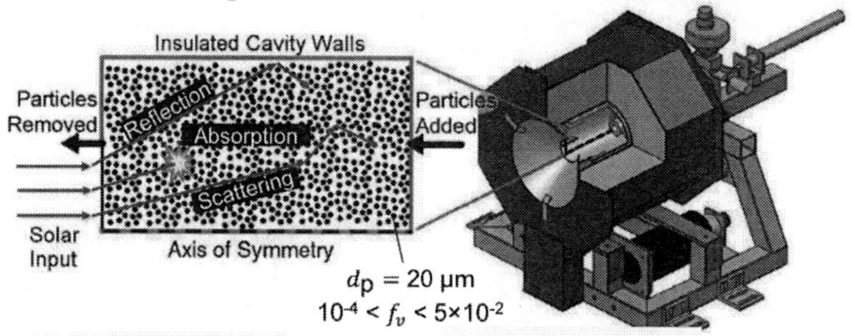

Conservation of Species (Reaction extent)

$$0 = -\nabla \cdot \left(\vec{G}''_{Co_3O_4}\right) - r'''$$

Particle motion

Rate of conversion of Co_3O_4 to CoO

Conservation of Energy (Temperature)

$$0 = -\nabla \cdot \left(-k_e \nabla T + \vec{G}''\hat{c}T\right) - q'''_{rad} - r'''\Delta H_r$$

Conduction

Particle motion

Volumetric Radiation (participating media)

Rate of energy consumption from reaction of Co_3O_4

Equations solved using the Finite Volume Method

VALPARAISO UNIVERSITY

Model Results: Impact of Feed Rate

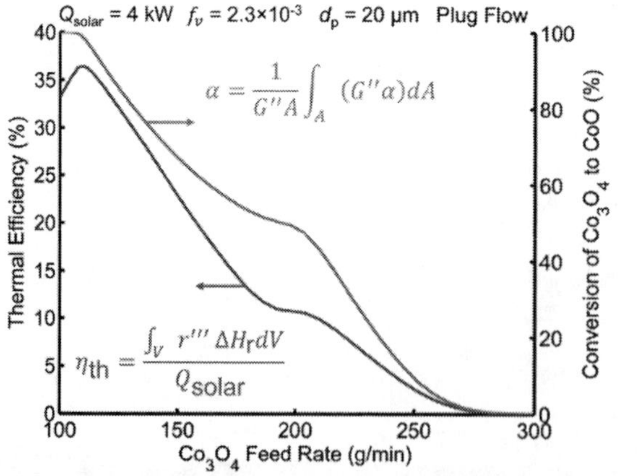

Maximum thermal efficiency achieved when the concentrated solar radiation is matched with the rate of the chemical reaction (α = 100%).

VALPARAISO UNIVERSITY

Reactor Testing

VALPARAISO UNIVERSITY

Reactor Operation

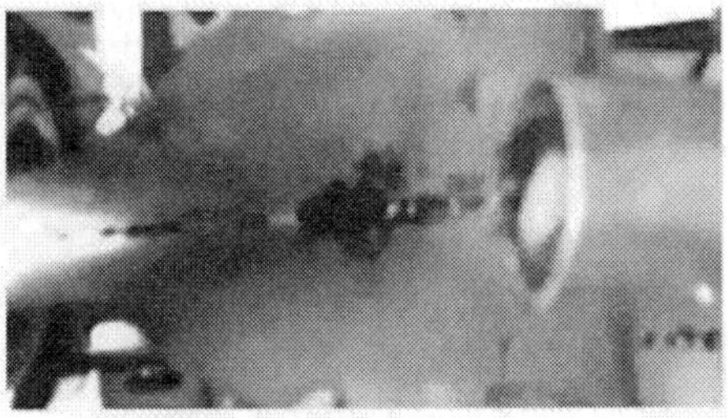

Reactor Inlet

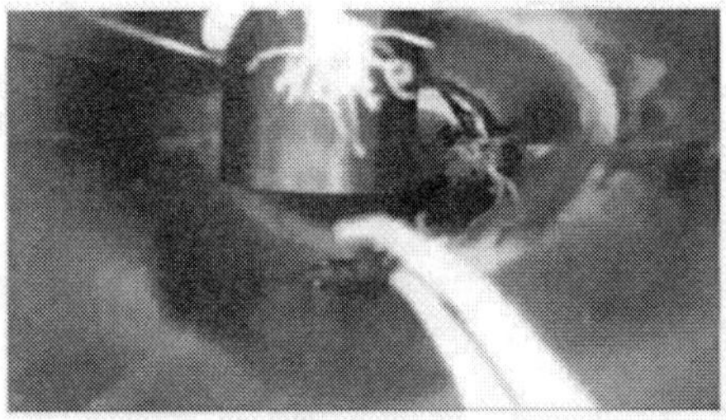

Reactor Outlet

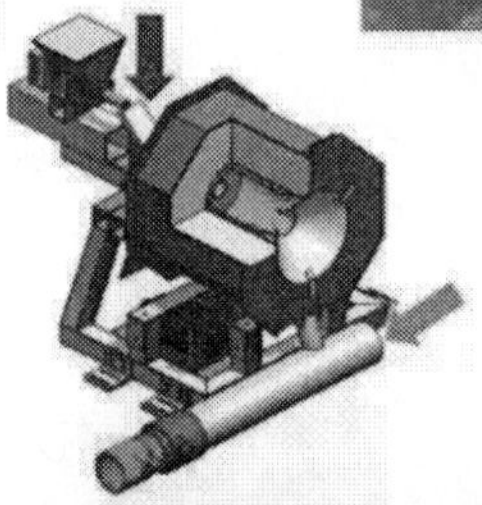

VALPARAISO UNIVERSITY

Reactor Test Results

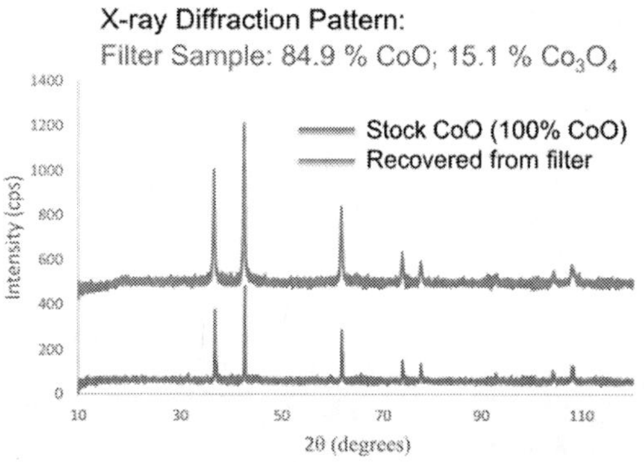

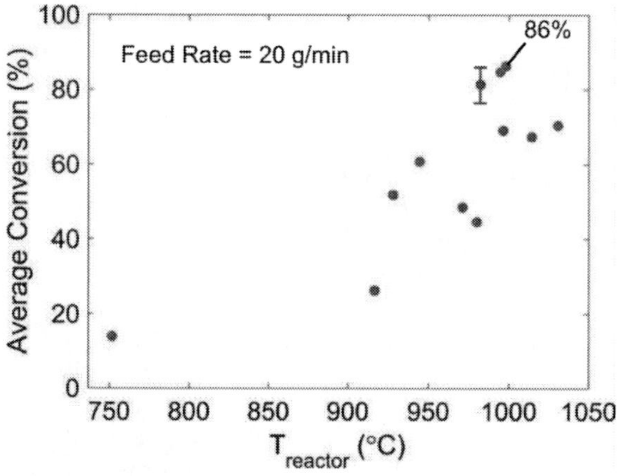

VALPARAISO
UNIVERSITY

Project Roadmap ($2.3 Million Grant)

VALPARAISO UNIVERSITY

Project Roadmap ($2.3 Million Grant)

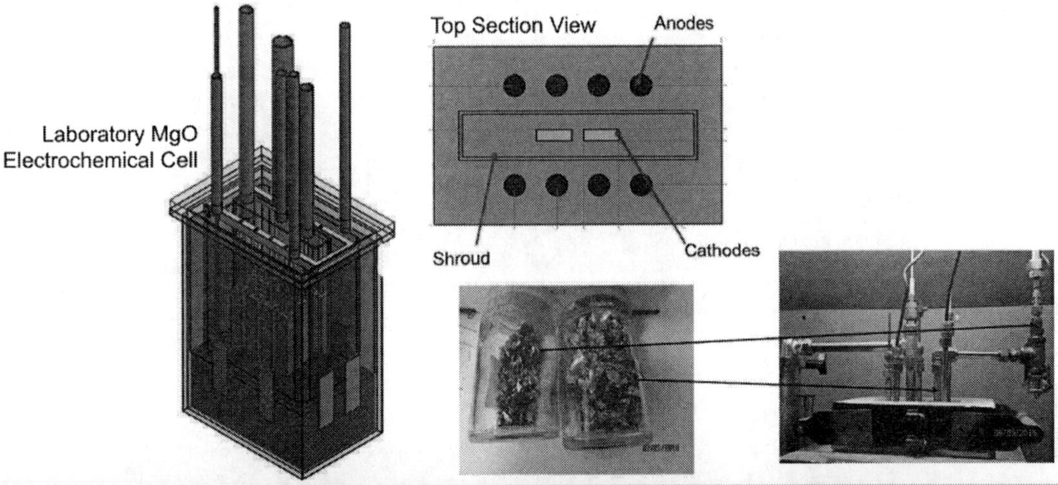

VALPARAISO UNIVERSITY

Claire Merrill

Fundamental Studies

VALPARAISO UNIVERSITY

Cathode Physics

Global Reaction: $Mg^{2+} + 2e^{-1} \leftrightarrow Mg\ (l)$

Reaction Kinetics: $j = j_o \left[1 \cdot \exp\left(\frac{(1-\alpha)F\eta}{RT}\right) - \frac{C_{Mg^{2+}}}{C^{\infty}_{Mg^{2+}}} \exp\left(-\frac{\alpha F\eta}{RT}\right) \right]$

Exchange Current Density (j_o); Transfer Coefficient (α)

Mass Transfer: $0 = \frac{\partial}{\partial t}\left(C_{Mg^{2+}}\right) - D_{Mg^{2+}} \nabla^2 C_{Mg^{2+}}$

Diffusion Coefficient ($D_{Mg^{2+}}$)

$\frac{-j}{nF} = D_{Mg^{2+}} \nabla C_{Mg^{2+}}$ } At the cathode-electrolyte interface (Zero flux at all other boundaries)

VALPARAISO UNIVERSITY

Results from 3D Finite Element Model

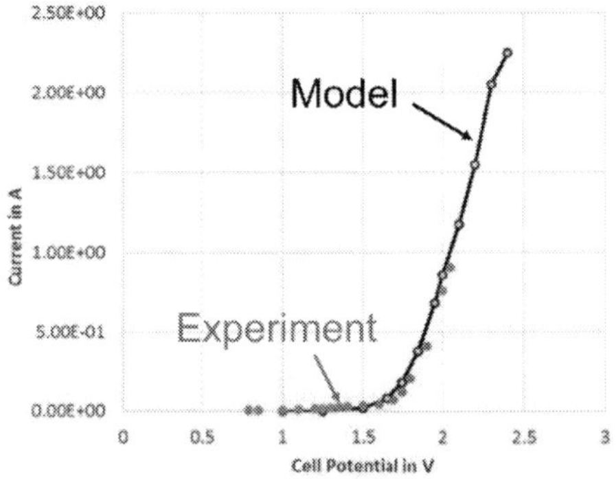

Simple Geometry

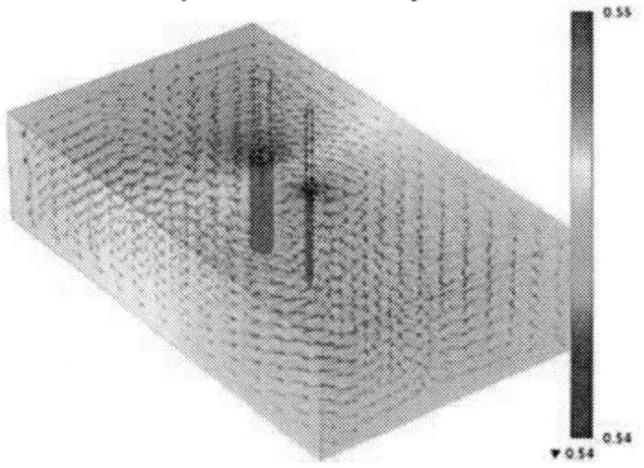

The cell is 13 cm x 7.5 cm x 2.54 cm

Anode = 4.0 cm^2 Cathode = 1.4 cm^2 Anode-Cathode Gap = 2 cm

VALPARAISO UNIVERSITY

Results from 3D Finite Element Model

Linear sweep voltammetry of Laboratory MgO Cell (20 A)

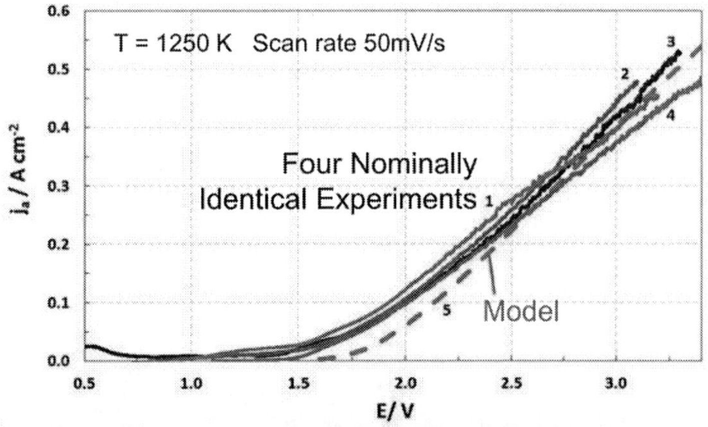

VALPARAISO UNIVERSITY

Towards Commercialization

VALPARAISO UNIVERSITY

Commercial Cell Concept

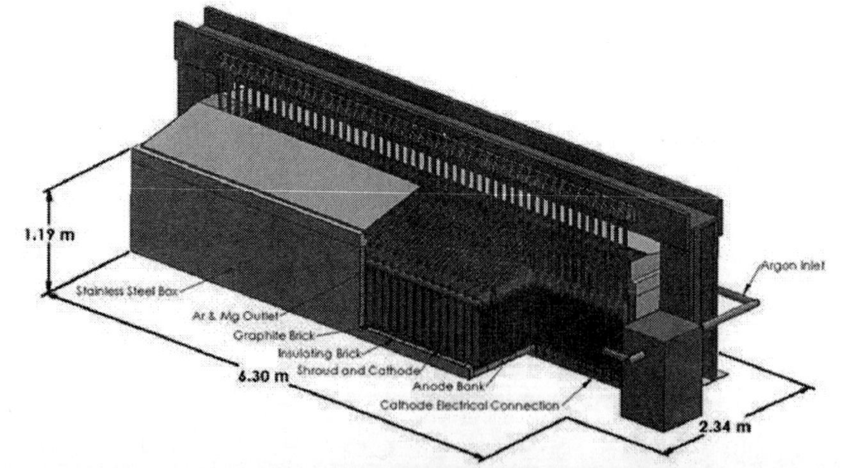

VALPARAISO UNIVERSITY

Finite Element Model of Commercial Cell Concept

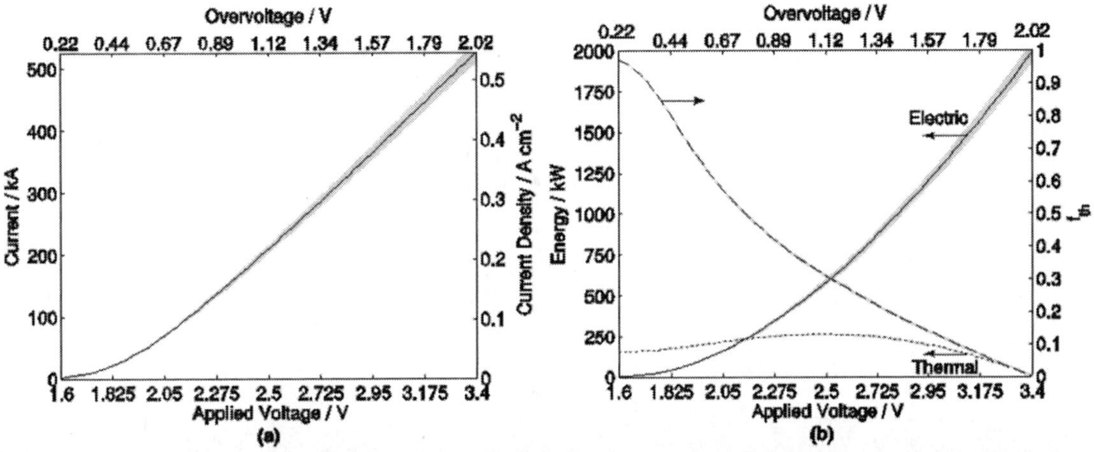

VALPARAISO UNIVERSITY

Visualizing the Solar-to-Commercial Cell Interface

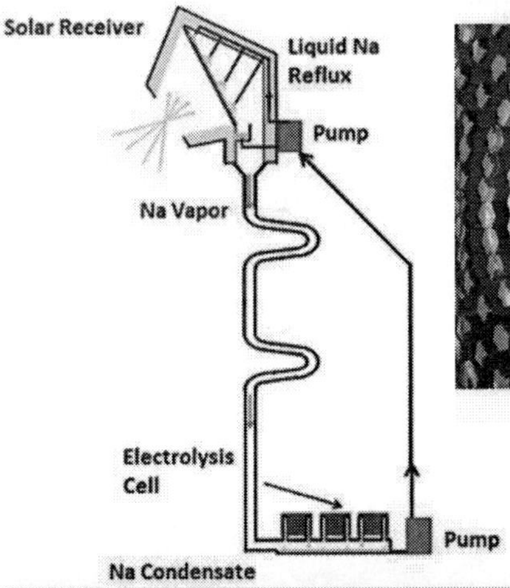

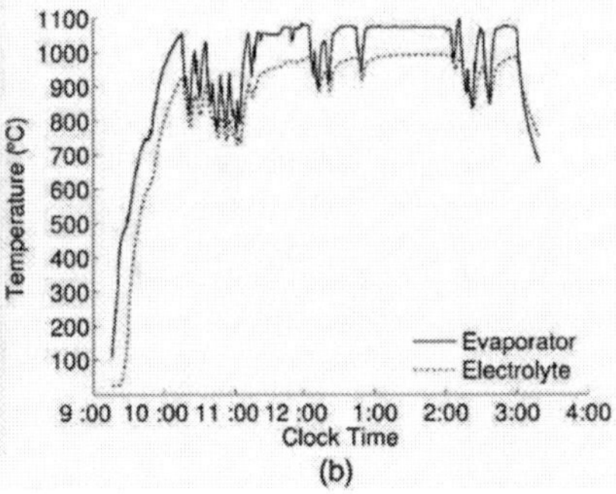

VALPARAISO UNIVERSITY

Growing Scientific Leaders

By Developing Deep Science and Engineering Knowledge

By Developing Communication Skills

By Developing the Passion to Serve Others

VALPARAISO UNIVERSITY

Undergraduate Student Co-authors

1) Leonard, N., Korenko, M., Larson, C., Blood, K., Venstrom, L.J., Nudehi, S., Duncan, S., Diver, R., Simko, F., Priscak, J. and Schoer, J., 2016, "The thermal electrolytic production of Mg from MgO: A discussion of the electrochemical reaction kinetics and requisite mass transport processes," *Chem. Eng. Sci.*, **148**, pp.155-169.
2) Palumbo, R., Korenko, M., Larson, C., Venstrom, L. J., Duncan, S., Nudehi, S., Schoer, J., Toberman, J., Prusinski, W., Johnson, D., Robinson, B., Barkely, S., Warren, K., Diver, R., Šimko, F., Boča, M., 2015, "Thermal Electrolytic Production of Mg from MgO: Reflections on Commercial Viability", *144th Annual Meeting & Exhibition of the Minerals, Metals & Materials Society TMS2015*, March 15-19, 2015, Orlando, FL.
3) Sheline, W., Matthews, L., Lindeke, N., Duncan, S., Palumbo, R., 2013, "An Exploratory Study of the Solar Thermal Electrolytic Production of Mg from MgO," *Energy*, **51**, pp. 163 – 170.
4) Palumbo, R., Diver, R.B., Larson, C., Coker, E.N., Miller, J.E., Guertin, J., Schoer, J., Meyer, M., Siegel, N.P., 2012, "Solar Thermal Decoupled Water Electrolysis Process I: Proof of concept," *Chem. Eng. Sci.*, **84**, pp. 372-380.
5) Schroeder, R., Matthews, L., Leatzow, D.,Kondratko, J., Will, J., Duncan, S., Sheline, W., Lindeke, N., Neves, P., and Palumbo, R., 2011, "Solar Thermal Electrolytic Process for the Production of Zn from ZnO: The Electrolysis of ZnO from 1275-1500 K," *J. Solar Energy Eng.*, **133**(4)
6) Krenzke, P., Krueger, K., Leonard, N., Duncan, G., and Palumbo, R. D., 2010, "A Solar Thermal Electrolytic Reactor for Studying the Production of Metals from Their Oxides," *J. Solar Energy Eng.*, **132**(3), pp. 34501-34505.
7) Venstrom, L., Krueger, K., Leonard, N., Tomlinson, B., Duncan, G., and Palumbo, R.D., 2009, "Solar Thermal Electrolytic Process for the Production of Zn from ZnO: An Ionic Conductivity Study," *J. Solar Energy Eng.*, **131**, pp. 1-9.
8) Freid, A.P., Johnson, P.K., Musella, M., M ller, R., Steinbrenner, J.E., and Palumbo, R.D., 2005, "Solar blind pyrometer temperature measurements in high temperature solar thermal reactors: a method for correcting the system-sensor cavity reflection error," *J. Sol. Energy Eng.*, **127**(1), pp.86-93.

VALPARAISO UNIVERSITY

Undergraduate Student Presentations

VALPARAISO UNIVERSITY

Student Outcomes: Graduate School

Graduate Schools Attended:

- Georgia Tech
- Penn State
- University of Florida
- University of Michigan
- University of Minnesota
- Harvard University
- University of Illinois at Urbana-Champaign
- Michigan State
- Stanford
- Vanderbilt
- Ohio State University
- University of Wisconsin
- Colorado State

64 percent enrolled in a graduate program (last 3 years)

4 NSF Graduate Research Fellowship winners and 4 honorable mentions (last 5 years)

VALPARAISO UNIVERSITY

PROF. SCOTT SANDERS

DEPARTMENT OF MECHANICAL ENGINEERING, UNIVERSITY OF WISCONSIN-MADISON

A sustainability challenge: communicating science with the general public at an emotional level without compromising the science

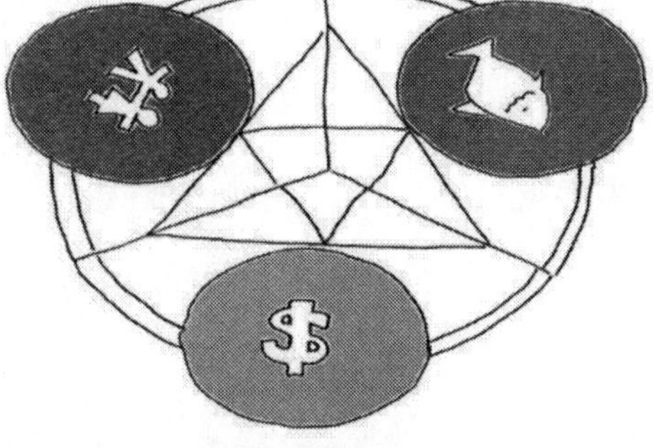

VALPARAISO UNIVERSITY

Thank you, Professor Edward Fletcher,
for helping us to Grow Future Scientific
Leaders with Concentrated Sunlight.
With Love, The VU Community
VALPARAISO
UNIVERSITY

495673 Metal Oxide Redox Materials for Solar Thermochemical Energy Storage

Abstract:

Thermochemical energy storage (TCES) systems are of great interest in concentrated solar power (CSP) applications. Storing sunlight as chemical energy during the day can enable power generation at night or during cloudy periods, effectively alleviating the inherent intermittency of solar sourced electricity. Metal oxides are among the most attractive TCES materials because they possess high-energy density and high reduction/oxidation temperatures (>800°C) suitable for driving high-efficiency thermodynamic power cycles. However, limitations involving high-temperature heat transfer, particle handling, sintering, and chemical performance of the metal oxide materials are still problematic in practical application.

Past works generally focus on evaluating pure metal oxide systems and investigate a select few mixed metal oxide systems in detail. In this work, we pre-selected over 100 high-potential mixed-metal oxide systems based on a comprehensive literature survey and then screened those materials for suitability in redox based high-temperature TCES using thermodynamic database software. The potential candidates were down-selected to the most promising contenders based on several selection criteria: high-temperature stability, large redox reaction enthalpy, and reaction temperatures of approximately 700-1100 Â°C for CSP applications. Ongoing and future work we will focus on experimental testing of the identified compositions to identify actual redox behaviours and to evaluate other thermochemical criteriaâ€"appropriate cycling physical stability, maintained chemical activity over many redox cycles, and acceptably fast reaction kinetics.

502849 SolPeD - solar per demand for power generation and fuel production

Abstract:

We are challenged to develop an energy conversion method, which could gradually replace fossil fuel combustion on a worldwide scale, eventually leading to a renewabledominant global energy supply. Whereas other renewable energy resources can have important supporting roles, the only energy sources available on earth in large enough quantities to replace fossil fuels are nuclear and solar.
The talk will describe a solar energy conversion process based on several complementing novel technologies, and aimed to enable large-scale utilization of solar energy for yearround, 24/7 power generation and clean fuel production at high efficiency and affordable costs.
The major technological challenges and proposed solutions will be discussed, and ongoing effort of commercializing the technology will be reported.

498174 On the Role of Thermochemical Energy Storage in Concentrating Solar Power

Abstract:

Concentrating solar power (CSP) with thermal energy storage has the potential for grid-scale dispatchable power generation. Thermochemical energy storage (TCES), that is, the reversible conversion of solar-thermal energy to chemical energy, has high energy density and low heat loss over long periods. Previously, several reaction types have been studied as candidates, including dissociation of ammonia, hydrides, carbonates, oxides, hydroxides and methane reforming [1]. Although considerable work has been done, past research has mostly focused on experimental studies and reactor design. The limited system-level studies that are available in the literature are for specific reactions [2] and, furthermore, consider TCES in isolation, neglecting its interactions with the solar field and power generation system [3].

Accordingly, in this work, we establish a framework to systematically analyze and compare candidate reactions for TCES systems [4]. First, we develop a general process for CSP plants employing TCES systems. A general TCES system includes five tasks: reaction, storage, separation, material handling and heat recovery. We then formulate an optimization-based model for the process, which optimizes the plant design towards the minimum levelized cost of electricity (LCOE). We use this model to study ammonia and methane systems.

We identify that, for fluid phase TCES systems, the main cost driver is gas storage using pressure vessels, accounting for 77% of the TCES cost. This is because, even if reaction products are compressed and stored at 550 bar, both ammonia and methane systems require 65,000 m^3 of gas storage at a vessel price of 8900 $\$/m^3$. The main energy driver is the electricity consumption for storage compression, which consumes 18% of the generated power. Adopting underground gas storage, where products are stored at lower pressure in much cheaper underground caverns, could significantly reduce the compression power consumption (from 18% to 3%) and improve the overall energy efficiency and thus cost. The methane system combined with underground storage is identified as the most cost-efficient alternative, achieving a 13% LCOE reduction over the current two-tank molten salt CSP plants.

Finally, we discuss the impact of the reaction temperature and operating pressure. The endothermic and exothermic reaction temperature determines not only the reaction conversion, but also the receiver and turbine efficiency. We find that both reaction temperature should be in the vicinity of the reaction turning temperature (at which the equilibrium constant K=1) and separation can increase the reaction conversion, allowing lower endothermic reaction temperature. The operating pressure, including process pressure, highest and lowest storage pressure, influences both the storage volume and compression work. For pressure vessel cases, the storage tank pressure varies in a wide range (150-550 bar), reducing the storage volume at the expense of more compression power consumption. When cheap underground storage is employed, a narrow storage pressure range (135-165 bar) leads to smaller electricity consumption and larger storage volume.
The proposed framework can be used to study a wide range of TCES systems, including solid-gas phase reactions, which will be the topic of future research.

References:

[1] P. Pardo, A. Deydier, Z. Anxionnaz-Minvielle, S. Rougé, M. Cabassud and P. Cognet, *Renew. Sustain. Energy Rev.*, 2014, **32**, 591–610.

[2] A. Luzzi, K. Lovegrove, E. Filippi, H. Fricker, M. Schmitz-goeb, M. Chandapillai and S. Kaneff, *Sol. Energy*, 1999, **66**, 91–101.

[3] C. Corgnale, B. Hardy, T. Motyka, R. Zidan, J. Teprovich and B. Peters, *Renew. Sustain. Energy Rev.*, 2014, 38, 821–833.

[4] X. Peng, T. W. Root and C.T. Maravelias, *Green. Chem*, 2017, DOI: 10.1039/c7gc00023e.

500484 Combined Thermochemical and Latent Heat Energy Storage for Low Temperature Residential Applications

Abstract:

Solar thermal energy has the potential to supply clean energy, but for space heating and cooking applications, peak usage times often are misaligned with peak solar hours. Solar thermochemical energy storage systems are one promising solution. Salt-hydrate reactions are well-suited to this purpose as they have low-decomposition temperatures and high energy densities. To enhance heat transfer, improve durability, and prevent conglomeration, a solid matrix material is used, typically made of porous substances such as activated carbon, vermiculite, or expanded graphite. Here, a matrix of expanded graphite reinforced with carbonized flour was used. Certain salt-hydrates have the potential to store both thermochemical and latent heat, and one such salt was investigated through TGA/DSC analysis. A functional eutectic was found at around 130°C and freezing/melting cycles were visible as well-defined peaks. The salt-matrix complex was also investigated through both TGA/DSC and realistic application testing. The success of this new chemistry sets the stage for a new dual-release approach to energy storage.

496055 Optical Losses from Bubble Evolution in Photo-Electrochemical Reactors

Abstract:

Solar fuels, such as the production of hydrogen by photo-electrolysis of water, have the potential to revolutionise the long term future of energy production as they can overcome the issue of intermittency. The potential of this technology can only be realised by scaling up current research and lab scale demonstrations and henceforth, design of photo-electrochemical (PEC) reactors will become increasingly important.

Successful PEC reactor design will minimise resistive losses in solution, ensure the separation of the photo-electrolysis products and permit maximal light absorption by the photo-electrode(s). Methodologies for simultaneously achieving these requirements have not been exhaustively studied and in particular a comprehensive quantification of optical losses in PEC is missing from the literature.

In this work we consider the optics of PEC reactor design and its ramifications, focusing on the optical losses of bubbles on the surface of the photo-electrode. Bubble evolution is a complex process of bubble nucleation, detachment and rise and not only detrimentally reduces the effective surface exposed to electrolyte but absorbs, reflects and scatters incoming photons.

We experimentally demonstrate this effect and quantify the optical losses due to bubble evolution using a transparent conductive oxide such as fluorine doped tin oxide (FTO) glass as an anode. The optical losses are correlated against current density (and hence gas evolution rate) and the ramifications of this discussed. From this relationship, a modified maximum Solar-To-Hydrogen efficiency for different semiconductors (Shockley-Queisser limit) is derived.

Opportunities for the elimination of optical losses from bubble evolution through bubble suppression are discussed where pressurisation and forced convection approaches are investigated. A multi-physics model is presented that demonstrates the operational requirements in order to a achieve bubble suppression in a novel membrane-less flow reactor design, which has been outlined in previous work. Experimental verification of bubble suppression through hydrodynamic control is given and compared to the theoretical results and the potential scalability of this technology explored.

497838 Transport Analysis of an Integrated Artificial Photosynthetic System for Direct Capture and Reduction of CO_2 from Air

Abstract:

The effect of anthropogenic emissions of CO_2 on the climate change is becoming progressively prevalent. The current concentration of CO_2 in the air is 409.11 ppm, which is increasing at a rate of ~1.9 ppm per year. There exist several technologies such as amine scrubbing, oxyfuel combustion, and calcium looping to capture CO_2 from flue gases, which are inherently energy intensive. However, there are only a few techniques such as moisture-swing adsorption that can capture CO_2 from the air but requires modulation of humidity and downstream separation. The captured CO_2 can be either sequestered at the storage site or chemically transformed into a usable product. Integrating CO_2 capture with reduction not only eliminates the need for additional transportation and storage but also increases the efficiency of CO_2 capture. Here we provide a conceptual design of an integrated system where an anion-exchange resin for CO_2 capture is directly connected to a photo/electrochemical cell for CO_2 reduction. The dry-side of the resin captures CO_2 from the air and produces a high concentration of HCO_3^- diffusing towards the wet-side of the resin, where HCO_3^- converts back to CO_2 at the attached cathode to produce fuels. The limiting current density of an integrated system as a function of resin properties, catalysts activity and selectivity, electrolyte pH, physical dimensions and relative humidity are simulated using COMSOL Multiphysics. The limiting current density increases with increasing the resin conductivity, and mass-transfer coefficient of CO_2 in air. The efficiency of such integrated system using air as well as flue gas will be discussed.

500306 Design of a 400 MW Carbon-Neutral, Coal-Fired Power Plant with Integration of Waste Heat and Solar Energy

Abstract:

Coal-fired power plants (CPP) consumes on an average of 221 tons h^{-1} of coal to generate 400 MW of electricity. The rate of emission of CO_2 from the state-of-the-art CPP operating at 60% efficiency is approximately 600 kg MW-h^{-1}, which is equivalent to 1.5 kmol s^{-1}. The amine-based scrubbers are known to effectively separate CO_2 from the flue gases which typically contains 12-15% CO_2. The captured CO_2 is then released using a regeneration process requiring an additional power of ~350 MW, which is typically supplemented by either burning natural gas or integrating with waste heat of ~250 MW from the CPP. Here we are proposing an electrochemical technique that directly coverts CO_2 absorbed in an amine-based solvent and water to either CO or HCOOH. This technique bypasses the need for regeneration process for amine-based solvent that saves ~350 MW of power, which can be used to conduct electrochemical reduction of CO_2. We found that continuous production of CO and HCOOH require lesser power as compared to other alcohols and hydrocarbons. For example, the minimum power required to produce either CO or HCOOH is ~585 MW (~380 MW is a thermodynamic limit) for 100% CO_2 Utilization Efficiency and Potential. The excess energy required for electrochemical conversion of CO_2 may be derived from the solar energy which requires a land area of 1.67 km^2 (38% of CPP land area), assuming a 20% effective efficiency of solar cells. Therefore, it seems plausible to support the energy requirement for electrochemical reduction of CO_2 using an integrated waste heat and solar energy management. The target performance of such system is ~1 A cm^{-2} of current density at ~2 V of cell voltage. Achieving such targets require development of a high surface area catalysts operating at total overpotential of <670 mV. This results in a carbon-neutral model that can be used as a basis for designing and adapting current and new conventional thermal power plants.

489902 The Energetics of Chemical Looping Processes for Post Combustion CO_2 Capture

Abstract:

The energetics of chemical looping processes for post combustion CO_2 capture

Hugo S. Caram[*], Ramesh Gupta[#], Fan Ni[*] and Hans Thomann[#]

(*) Department of Chemical and Biomolecular Engineering, Lehigh University, Bethlehem, PA

(#)ExxonMobil Research and Engineering Inc., Annandale, NJ

A number of processes have been discussed where the CO_2 is removed from flue gases by chemical reaction with a solid sorbent. While calcium looping is the most celebrated, many others including carbonate-bicarbonate, reactions with mixed oxides, treated hydrotalcites and organic impregnated solids, such as MOFS, have been proposed. The equivalent work requirement, that is the most important variable, power that will be drawn from the power plant are usually only examined when a detailed process design and pilot testing has been completed.

In this work, a simple thermodynamic tool is proposed that will allow determination of the

minimum equivalent work needed by the process. The tool only requires thermochemical properties at room temperature and evaluates major thermal and irreversible losses. It can be used to determine the effect of process parameters, equipment configuration (such as fluidized vs. fixed or moving bed operation), steam usage and other variables. Applications to some of the existing processes, where the method could apply, will be presented.

493365 A Technoeconomic Evaluation of Carbon Dioxide (CO_2) Electroreduction: Results and Implications

Abstract:

The electroreduction of carbon dioxide (CO_2) to value added C_1-C_2 chemicals (such as formic acid, carbon monoxide, methanol, ethylene, and ethanol) has been proposed as a possible technological solution that can utilize excess anthropogenic CO_2 emissions, thereby providing a sustainable pathway for closing the CO_2 emission cycle.[1, 2] However, even after a few decades of research, key questions regarding the technoeconomic feasibility of the process remain unanswered. These include: (i) what products of CO_2 electroreduction are the most economically viable to produce; and (ii) what performance targets such as maximum operating cell potential (V_{max}), minimum operating current density (j_{min}), Faradaic efficiency (FE), and catalyst durability (t_{catdur}), are required for an economically viable process.

In this talk, we will present results from a gross-margin based economic model that we recently developed to evaluate the technoeconomic feasibility of the CO_2 electroreduction process.[3] Using the model, we were able to quantitatively estimate the operating parameters (V_{max}, j_{min}, FE, t_{catdur}) required for an economically viable process of CO_2 electroreduction. A comparison of the operating parameters required for economic viability with the experimentally achievable performance of the state of the art electrochemical systems indicate that formic acid and carbon monoxide are the most economically viable CO_2 electroreduction products, and the process can be cost competitive with the existing industrial production methods for the same. However, to improve the economic viability of other products (such as ethanol, ethylene), alternate reaction chemistry/process designs with a lower electrical energy requirement need to be developed. This talk will cover some of our recent advances with respect to developing such alternate process design strategies as well.

References:

[1] A.M. Appel, J.E. Bercaw, A.B. Bocarsly, H. Dobbek, D.L. DuBois, M. Dupuis, J.G. Ferry, E. Fujita, R. Hille, P.J.A. Kenis, C.A. Kerfeld, R.H. Morris, C.H.F. Peden, A.R. Portis, S.W. Ragsdale, T.B. Rauchfuss, J.N.H. Reek, L.C. Seefeldt, R.K. Thauer, G.L. Waldrop, *Chem. Rev.* 2013, 113, 6621-6658.

[2] H.R.M. Jhong, S. Ma, P.J.A. Kenis, *Curr. Opin. Chem. Eng.* 2013, 2, 191-199.

[3] S. Verma, B. Kim, H.R.M. Jhong, S. Ma, P.J.A. Kenis, *ChemSusChem* 2016, 9, 1972-1979.

496476 Methanol Partial Oxidation on Silver in a Fixed Bed of Raschig Rings: An Integrated CFD with Microkinetics Study

Abstract:

Methanol partial oxidation on silver in a fixed bed of Raschig rings:

An integrated CFD with microkinetics study

Behnam Partopour, Anthony G. Dixon

Department of Chemical Engineering, Worcester Polytechnic Institute,

Worcester, MA, USA, 01609

Formaldehyde is extensively used in the chemical industry both as an intermediate and as a final product. Fixed bed partial oxidation of methanol over a silver catalyst is one of the main production processes for formaldehyde. The silver catalyst is usually placed on cylindrical or ring shaped pellets. The reaction is highly exothermic with carbon dioxide as a by-product. Andreasen et al., proposed a microkinetic model to explain the reaction mechanism which was validated by experimental data[1]. It is desired to study such detailed kinetics under realistic reacting conditions to understand the interaction of the fluid flow and transport with the kinetics, and how different parameters impact the overall performance of the fixed bed.

In this work we have coupled the microkinetics model with resolved particle fixed bed CFD simulations using a spline mapping method that we have recently developed[2]. Then the code is implemented for a packed bed geometry of 250 Raschig rings with tube to particle diameter ratio equal to five (i.e. N= 5 and d_p=8 mm). Previous resolved particle studies were all conducted in a packed bed of spherical particles[3], and this is the first time such a simulation is done for bed of Raschig rings.

The study is separated into two parts: 1- effect of surface species and operating conditions on local selectivity, 2- comparison between packed bed of Raschig rings and a packed bed of spherical particles.

Preliminary results show that the temperature gradient is dominantly controlling the reaction and species profiles in the bed. Since the temperature profile is correlated to the fixed bed configuration it could be shown that the species profile in the bed is affected by fixed bed configuration and so is the selectivity. Furthermore, the comparison between spheres and rings shows that due to higher contact area between fluid flow and catalyst surface, the local selectivity is improved in the case of Raschig rings. The results show that particles with higher surface to volume ratio, and thus higher rates of heat removal, are cooler, which prevents the further oxidation of the products. Furthermore, in such particles reactant species diffuse faster to the particle center which also helps to improve the local selectivity.

References

1- Andreasen, A., Lynggaard, H, Stegelmann, C, Stoltze, P. A microkinetic model of the methanol oxidation over silver. Surf. Sci. 2003;544(1):5-23.

2- Partopour, B., Dixon, AG. Computationally efficient incorporation of microkinetics into resolved-particle CFD simulations of fixed-bed reactors." Comp. Chem. Eng. 2016; 88:126-134.

2- Maffei, T., Gentile, G., Rebughini, S., Bracconi, M., Manelli, F., Lipp, S., Cuoci, A., Maestri, M. A Multiregion operator-splitting CFD approach for coupling microkinetic modeling with internal porous transport in heterogeneous catalytic reactors. Chem. Eng. J. 2016; 283:1392.

497523 Application of Recent Understanding of Intra-Molecular Forces and Micro-Hydrodynamics to the Bubble Dynamics and Gas-Liquid Transfer in a Downward-Flow Microbubble Column

Abstract:

Bioreactors have recently become a technology of interest for conversion of hydrocarbon gases to liquid fuels. Economic viability of this process requires exceptionally low reactor costs and high production rates. Therefore development of ambient pressure bioreactors with a fast gas-liquid mass transfer coefficient (k_La) and low energy consumption is desired. One strategy to achieve this goal is to increase interfacial area density, a, to >3000 m^2/m^3, with minimal use of energy. We compare published reactor technology with our own measurements from a new reactor design that utilizes downward-bubble flow away from an array of micro-jets (Li 2014). This advanced design achieves more efficient k_La per power expenditure while operating at higher k_La. We present measurements from four reactor heights (height-to-diameter ratios of 12, 9, 6 and 3) of k_L, total interfacial area a, liquid residence time distribution, energy consumption, turbulent hydrodynamics, bubble breakup size, bubble size distribution, surface tension, . A physical model for predicting k_L and is validated, and suggests that k_L is governed by different hydrodynamics at different locations below the micro-jet array: 1) "entrance effects" due to Higbie (1935) penetration dominate at short distances, 2) turbulence (Calderbank & Moo-Young 1961) dominates at intermediate distances, and finally 3) terminal rise velocity (Clift 1978) dominates at large distances. Recent advances in the understanding of molecular and hydrodynamic forces are applied to our data, including the effects of sodium dodecyl sulfate (SDS) and potassium chloride (KCl). We find that surface tension is not the leading physical force to control bubble breakage and coalescence rate. We use recent literature to make calculations of the leading intra-molecular and hydrodynamic forces. The experimental behavior of the bubbles is well explained by the sum of molecular and hydrodynamic forces. Improvements are suggested for predictions of bubble size in gas-liquid dispersions in agitated vessels, which classically assume that surface tension is the connection between molecular additives and macroscopic bubble behavior (Akita Yoshida 1973, Liao 2010).

References:

Li, X. (2014), United States Patent 20140212937A1, System and method for improved gas dissolution, edited by L. N. Z. Limited.

Higbie, R. 1935 The rate of absorption of a pure gas into a still liquid during short periods of exposure. Transactions of the American Institute of Chemical Engineers. 31, 365.

Calderbank, P. H. & Moo-Young, M. B. 1961 The continuous phase heat and mass transfer properties of dispersion. Chem. Engng Sci. 16, 39-54.

Clift, A., Grace, J. R. & Weber, M. E. 1978 Bubbles, Drops, and Particles. Academic Press, Inc. New York.

Akita, K. & Yoshida, F. 1973 Gas Holdup and Volumetric Mass Transfer Coefficient in Bubble Columns Ind. Eng. Chem. Process Des. Dev.12 (1), pp 76–80.

Liao, Y. & Lucas, D. 2010 A literature review on mechanisms and models for the coalescence process of fluid particles. Chemical Engineering Science. 65(10), pp 2851–2864.

496827 Experimental and CFD Simulation Study of Bubble Column Equipped with a Bundle of Heat Exchanging Tubes (Internals) for Fischer-Tropsch (FT) Synthesis

Abstract:

Experimental and CFD Simulation Study of Bubble Column Equipped with a Bundle of Heat Exchanging Tubes (Internals) for Fischer-Tropsch (FT) Synthesis

Abbas J. Sultan, Laith S. Sabri, Hayder Al-Naseri, Muthanna H. Al-Dahhan†

†Multiphase Reactors Engineering and Applications Laboratory (mReal)

† Department of Chemical and Biochemical Engineering, Missouri University of Science and Technology, Rolla, MO 65409-1230. USA

Abstract

The successful design and scale up of the bubble/slurry bubble column reactor are crucially dependent on the proper understanding of their hydrodynamics. Among others hydrodynamics parameters, knowledge of the gas holdup distribution and 3D liquid velocity factors are necessary to design and scale-up of such reactors. Although a large number of published studies on the subject of the bubble column without obstacles. Unfortunately, there is no study in literature accounts for the influence of the vertical internals), which is mimicked to the heat exchanger tubes in the Fischer-Tropsch synthesis (cover 25% of the total cross-sectional area of the column), on the gas holdup distribution and 3D liquid velocity at high superficial gas velocities. Therefore, this study aims for the first time to experimentally investigate and quantify the impact of the dense vertical heat exchanging tubes (internals) size on the time-averaged cross-sectional gas holdup and 3D liquid velocity at the wide range of superficial gas velocities by using gamma-ray computed tomography (CT) and Radioactive Particle Tracking (RPT) Techniques. The second part of this study is to develop 3D Computational Fluid Dynamics (CFD) simulation for bubble column with vertical internals and validate against the experimental data. Experimental and Simulation were performed in 6-inch bubble column with or without internals for air -water system. In this study, two sizes of Plexiglas® internals, covering 25% of the total cross-sectional area (CSA) of a column that simulates those used in Fischer-Tropsch synthesis, were investigated within a wide range of superficial gas velocities (0.05 to 0.45 m/s). The gained knowledge and findings are practical interest in the design, scale-up, and optimization of bubble/slurry bubble columns equipped with vertical internals. Furthermore, this work provided benchmark data to validate and developed the CFD simulation to be able to reproduce hydrodynamic parameters at a different size of columns and configuration of vertical internals as well as operating conditions. At the conference, the results and finding will be presented and discussed.

501583 Numerical Study of Hydrogen Production Via High-Temperature and Low-Temperature Water-Gas Shift Reactors' System: The Multi-Scale (Pellet-Reactor Scale) Modeling Approach and Simulation

Abstract:

The primary objective of this work is to develop an advanced, and detailed mathematical model of a packed bed reactors (PBRs) carrying out high-temperature and low-temperature water gas shift reactions (WGSRs) for the hydrogen production. In industry, the water gas shift reactors consist of two different temperature stages, the high-temperature shift reactor (HTSR) and the low-temperature shift reactor (LTSR) with a cooling process between them is a recommended for the operation of hydrogen generation. Therefore, detailed and advanced numerical simulations on the HTSR and the LTSR in series are performed to predict the behavior of hydrogen production system. After completing a single-pellet, non-isothermal, steady-state simulation, we couple our model with a non-isothermal (adiabatic), steady-state packed-bed reactor model to form a hybrid multi-scale reactor model. The velocity, temperature and species' concentration profiles along both the reactor length and the pellet radius are captured by using rigorously defined momentum, energy, and species transport models, accounting for the physical mechanisms involved in the system such as convection, conduction, and reaction-diffusion. This effort provides a detailed numerical insight into the transport phenomena with multiscale studies in porous media (pellet and reactor scales). The rigorous Maxwell-Stefan Model is applied on the reactor scale to account mass diffusion fluxes. On the other hand, Dusty Gas Model is considered to describe mass diffusion fluxes for the single pellet scale. Studies that include a broad range of the operating conditions and parameters (e.g. reactor temperature, reactor size (weight of catalyst), the inlet gas composition, and the inlet gas flow rate) are carried out in this paper, in order to investigate the upper and lower limit conditions' effects on the results.

A Numerical Model of Ambient Air Vaporizer (AAV) and Its Validation with Pilot Scale Data

2017. 11. 01

Yongkyu Lee

School of Chemical and Biological Engineering
Seoul National University

Table of Contents

Introduction

❖ Liquefied natural gas (LNG) re-gasification process

✓ Open Rack Vaporizer (ORV)

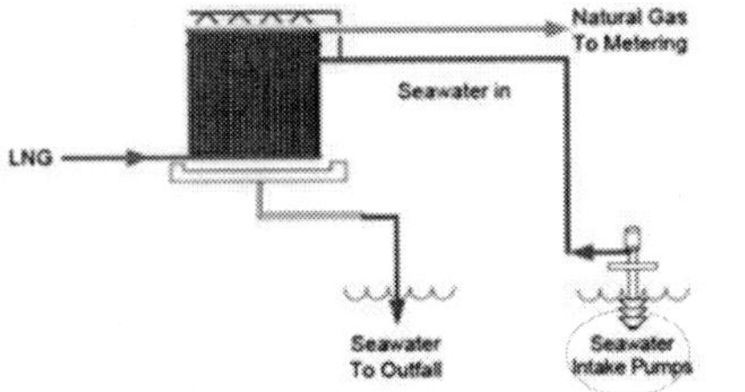

✓ Submerged Combustion Vaporizer (SCV)

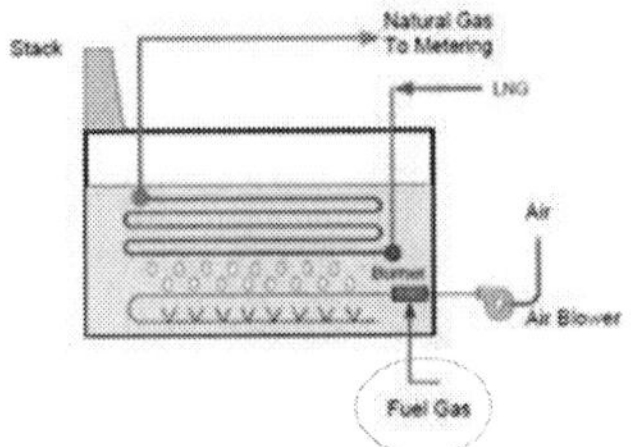

✓ Intermediate Fluid Vaporizer (IFV)

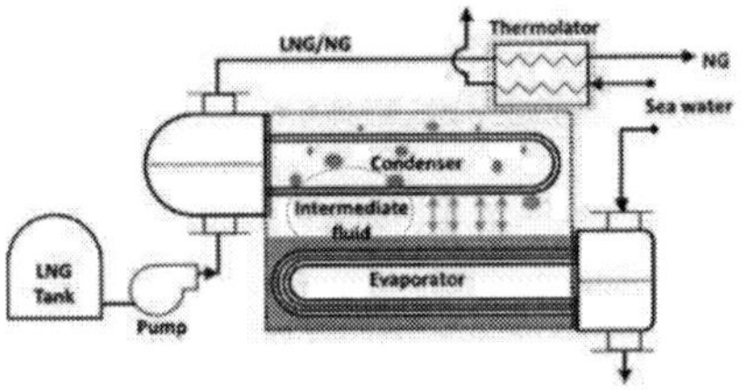

✓ Ambient Air Vaporizer (AAV)

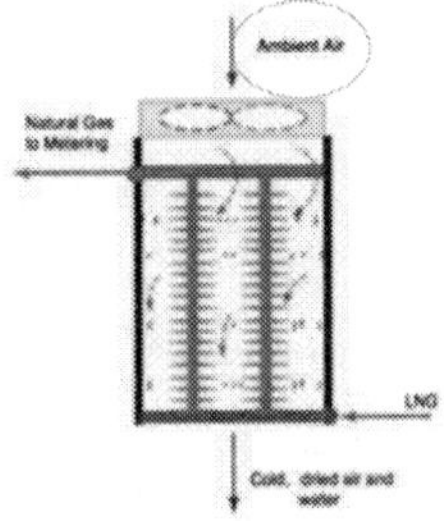

School of Chemical & Biological Engineering
Seoul National University

Introduction

❖ Ambient Air Vaporizer (AAV)

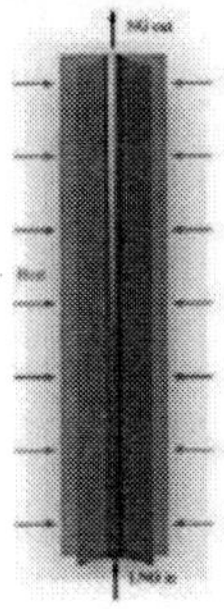

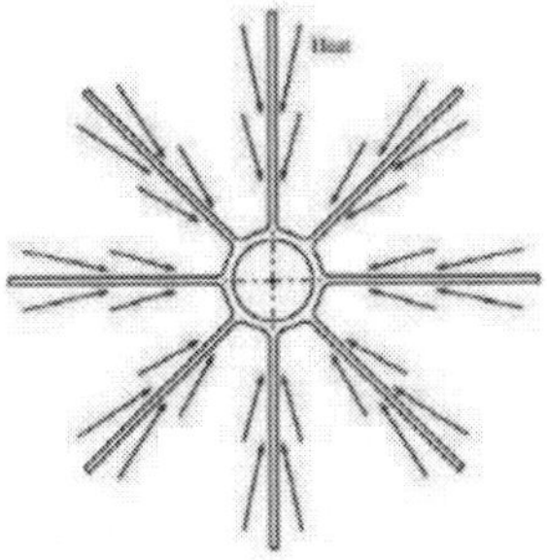

❖ Frost Formation

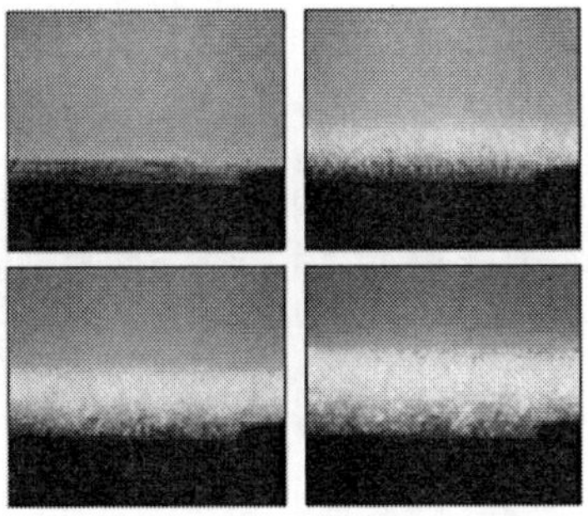

✓ It is critical to determine the duration for which the vaporizer is likely to be oprated with the NG of the outlet maintained at a stable temperature

✓ Predicting the frost accumulation during operation and heat transfer performance of the AAV over time

School of Chemical & Biological Engineering
Seoul National University

Introduction

❖ Ambient Air Vaporizer (AAV)

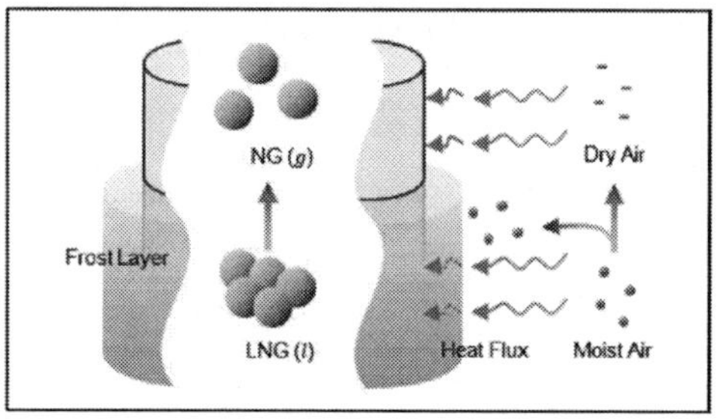

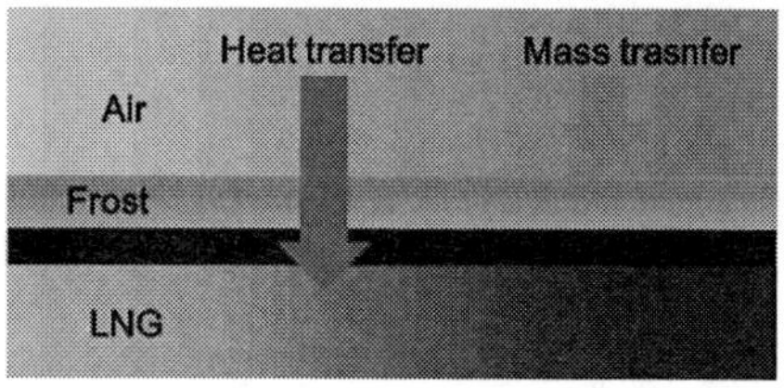

❖ Factors that affect thermal performance

✓ Air
- Velocity
- Temperature
- Humidity

✓ LNG
- Flowrate
- Temperature

✓ Frost
- Density
- Thickness

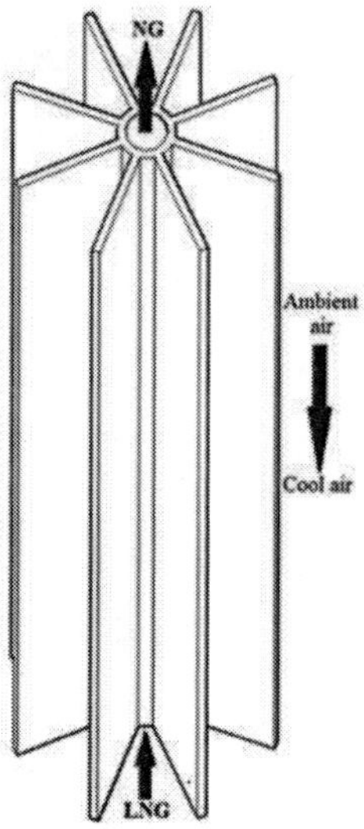

School of Chemical & Biological Engineering
Seoul National University

Experimental Measurement

❖ AAV used in experiment

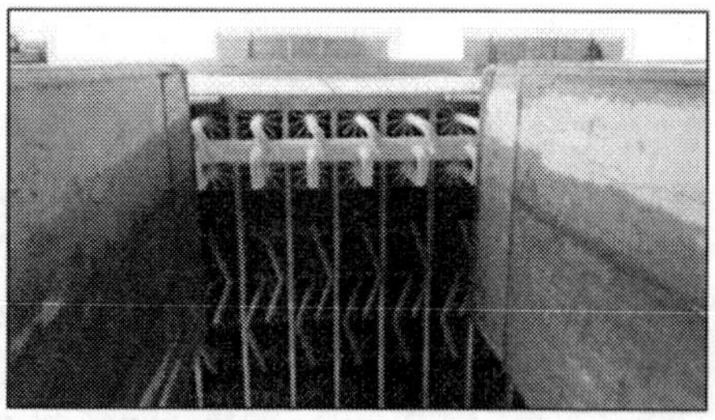

✓ AAV Capacity : 10 ton/h

✓ Fan Capacity : 4200 Nm³/min

✓ LNG inlet temperature : -158 °C

✓ Operation Pressure : 7.55 Mpa.g.

✓ Number of tubes : 24

✓ Number of fins on each tube : 12

✓ Length of tube : 72 m

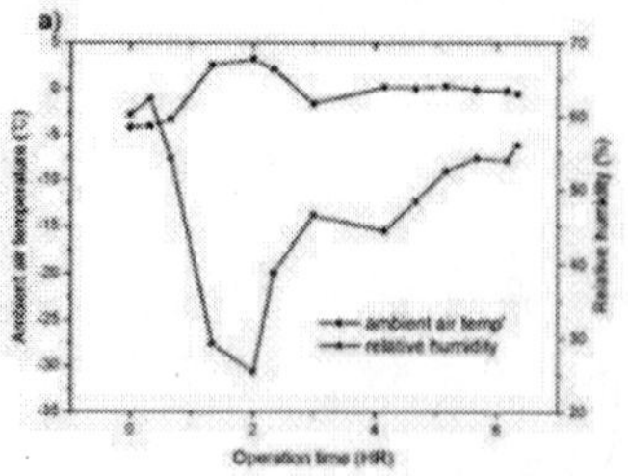

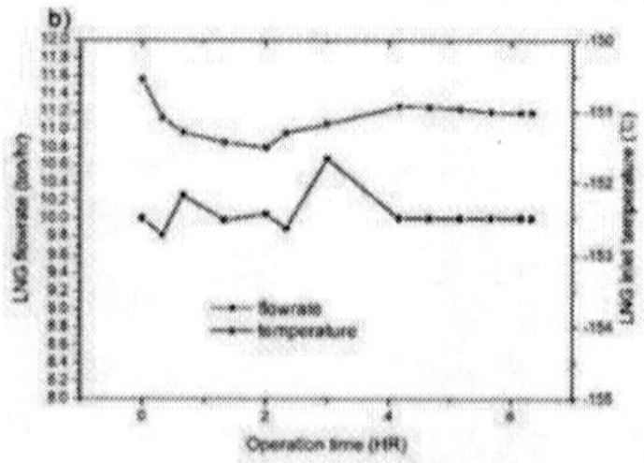

School of Chemical & Biological Engineering
Seoul National University

Numerical Model

❖ The schematic of the AVV and frosting process

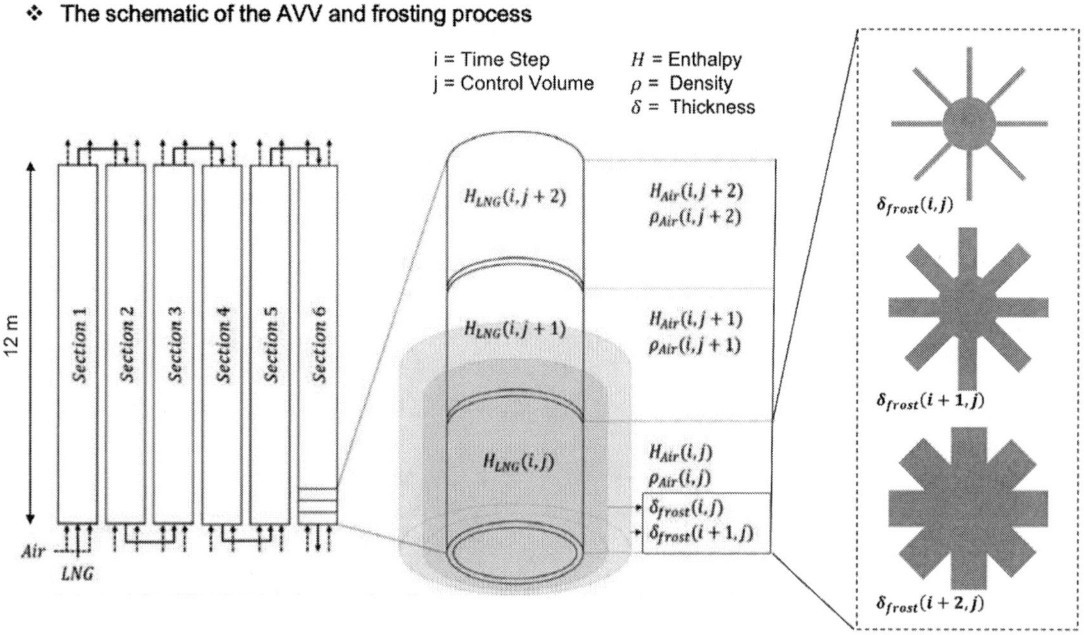

School of Chemical & Biological Engineering
Seoul National University

Numerical Model

❖ Flowchart of numerical model

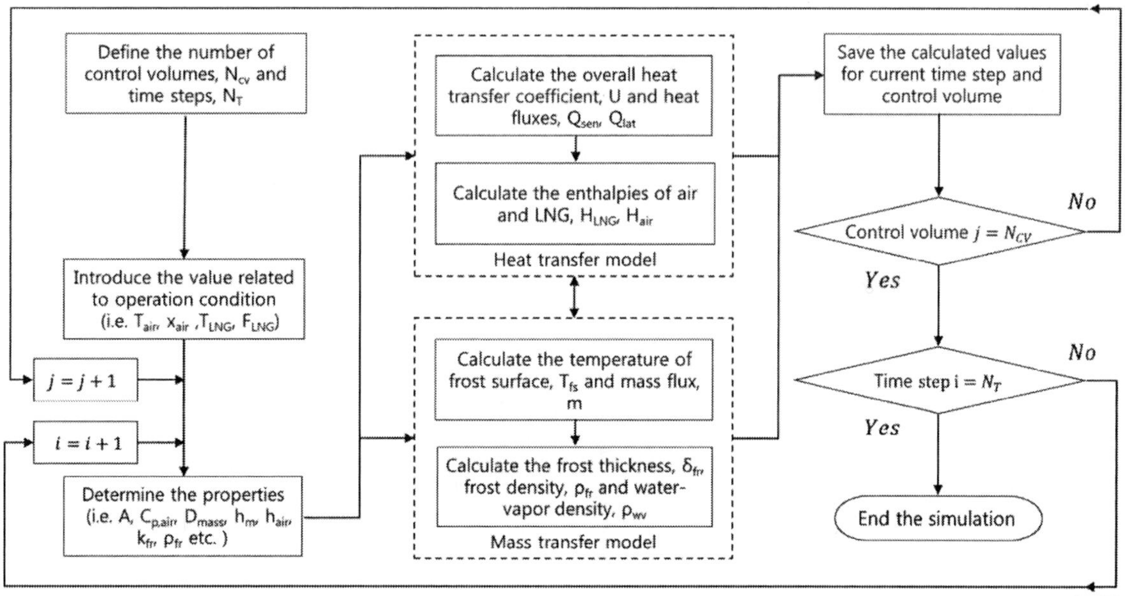

School of Chemical & Biological Engineering
Seoul National University

Result & Discussion

❖ Validation with pilot-scale experimental data

✓ The mean absolute error (MAE) between the pilot-scale data and model-predicted values was 5.5%.

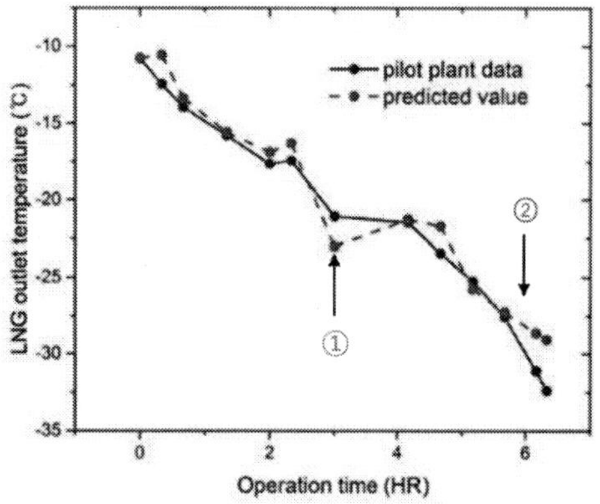

✓ ① : the LNG flowrate increased by 7% compared to the initial value, and the residence time of fluid become shorter, which reduces heat transfer efficiency.

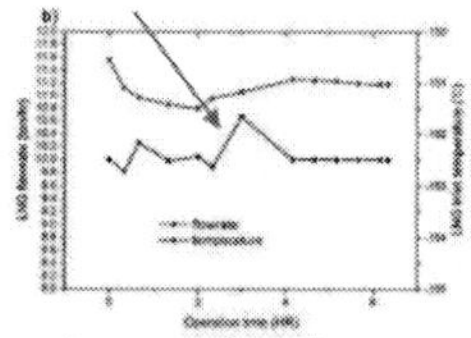

✓ ② : The space between the tubes arranged in parallel in the vaporizer is constant, and the frost growth clogs the space.

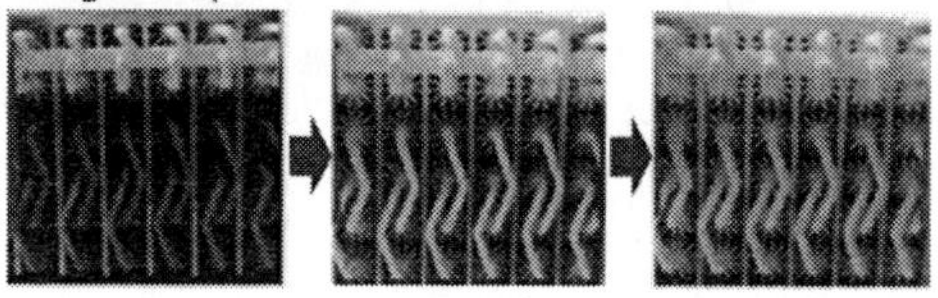

School of Chemical & Biological Engineering
Seoul National University

Result & Discussion

❖ Validation with pilot-scale experimental data

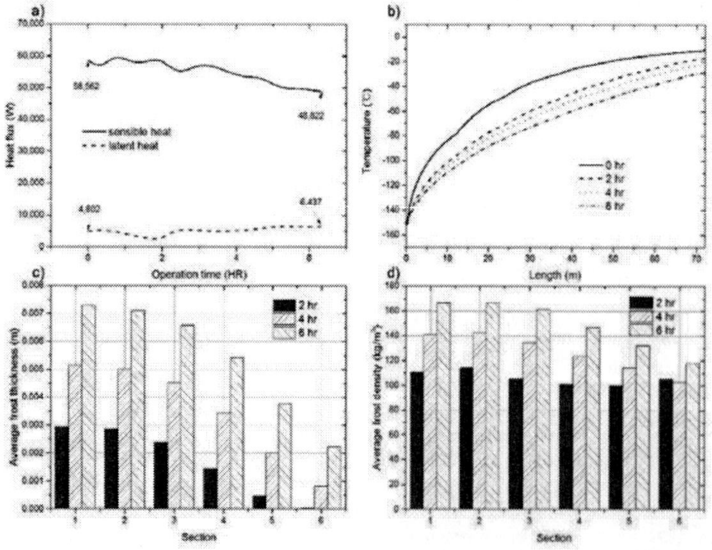

✓ a) sensible heat contributes 12 times more than latent heat to the total heat flux.

✓ b) As frost accumulates over time, the temperature of LNG in the vicinity of the outlet reduces to approximately -35 ℃ after 6 h of operation.

✓ c) It is observed that LNG is already vaporized in the preceding section as frost is not accumulated at the fifth and sixth sections after 2 h.

✓ d) The density of water vapor on frost surface at the rear portion of the vaporizer is higher than the front density and mass transfer

School of Chemical & Biological Engineering
Seoul National University

Result & Discussion

❖ Sensitivity Analysis

✓ Relative Humidity

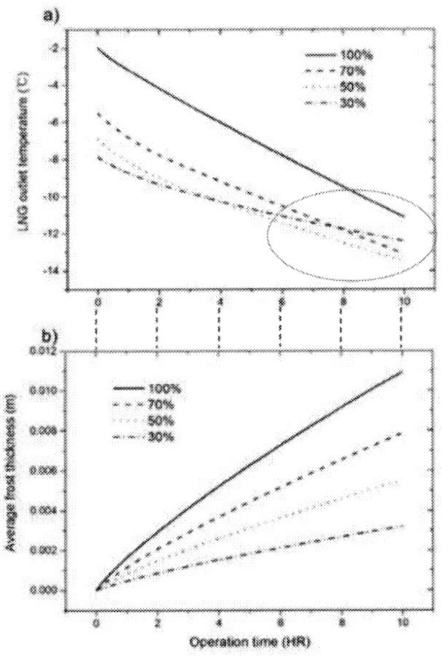

✓ Ambient Air Temp.

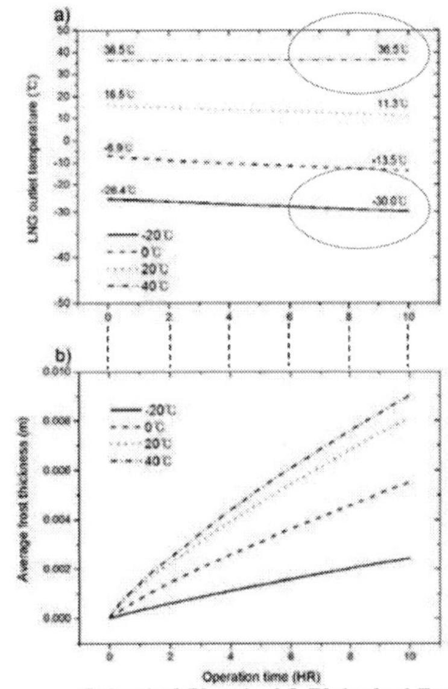

School of Chemical & Biological Engineering
Seoul National University

Result & Discussion

❖ Sensitivity Analysis

✓ Velocity of Air

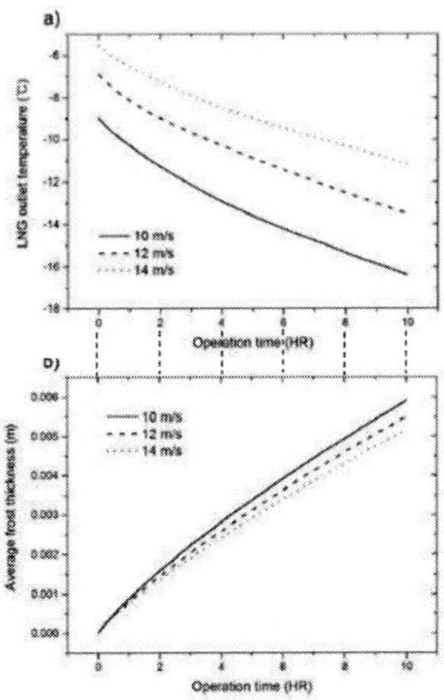

f Chemical & Biological Engineering
Seoul National University

Conclusion

❖ Conclusion

- ✓ Proposing a numerical model to predict the dynamic performance of the vaporizer by analyzing heat & mass transfer phenomena.
- ✓ The model is validated using the experimental data of a pilot-scale AAV.
- ✓ Sensitivity analysis for key factors such as air velocity, temp. and humidity.

❖ Limitation

- ✓ In pilot-scale units, there are uncontrollable variables such as geothermal and solar heat.
- ✓ The air stream causes physical movement of frost.
- ✓ It is challenging to consider the influence of frost scattering and fog during operation.

Thank You

Q&A

School of Chemical & Biological Engineering
Seoul National University

500177 Fluorescence Based Detection of Polychlorinated Biphenyls in Water

Abstract:

FLUORESCENCE BASED DETECTION OF POLYCHLORINATED BIPHENYLS IN WATER

Irfan Ahmad, J. Zach Hilt and Thomas D. Dziubla

177 F. Paul Anderson Tower, Department of Chemical and Materials Engineering, University of Kentucky, Lexington, KY 40506-0046

irfan.ahmad@uky.edu , thomas.dziubla@uky.edu

Even at very low ppb concentrations, the exposure to the halogenated persistent organic pollutants (POPs), such as polychlorinated biphenyls (PCBs), has been linked to numerous inflammatory diseases, including diabetes, cancer and lowered immune response. As a result, ready detection and sensing of PCBs in water is a critical need, yet is also highly challenging because of its low concentration and highly hydrophobic structure. While the gold standard remains gas chromatographyâ€"mass spectrometry (GC-MS), it requires complex sample preparation, is highly time consuming, and requires careful use of internal standards that increases its overall cost. As such, there is still need to develop a robust sensing technique to overcome these challenges. In this work, we evaluate a method for the sensing of organic pollutants using a hydrophobic fluorescent probe, benzopyrene (BP). As many organic pollutants, including PCBs are hydrophobic, BP fluorescence can be increased with trace concentrations of organic pollutants in water. BP forms hydrophobic complexes with organic pollutants that led to successful detection of pollutants including PCB-126, PCB-153, biphenyl and humic acid in water. Furthermore, this fluorophore can be absorbed onto the polymer microparticles (MPs) for advantages such as ease of use, regeneration of the detection system and reducing the overall cost of the method. These fluorophore-MPs attract organic pollutants in water to form hydrophobic complexes on the MPs and results in increase in fluorescent intensity. Polymer is bulk polymerized into a film using free radical polymerization method and then the film is cryogrind into microparticles. Moreover, the absorption and release of fluorescent molecule to and from the MPs has been studied using fluorescent kinetics technique to show the fast absorption and its stability in the water system. Particles were characterized using electron microscopy and their size was measured using light scattering technique. It was demonstrated that the developed material is effectively sensing PCBs in water.

492557 Soil Amendments Used during Remediation of Contaminated Sites Enhance Plant Growth but Increase Mobility of Chrysotile Fibers in One Case

Abstract:

The ability of plants to stabilize soil and reduce erosion, a process known as phytostabilization, can be critical in remediating or restoring abandoned surface mines or other sites contaminated with pollutants such as heavy metals. Nutrient-rich soil amendments are often applied to the soil surface to improve plant growth, hasten the development of vegetation, and reduce soil erosion and mobility of contaminants.

We were interested in whether asbestos fibers, as a contaminant, move in soil and whether phytostabilization is relevant to preventing their movement. We set up a greenhouse experiment in which we grew two crop species, *Brassica juncea* and *Sorghum bicolor*, in Cone-tainers^{TM} containing either pure sand or sand mixed with chrysotile fibers in a 1:15 ratio by weight of chrysotile:sand, layered over pure sand. The chrysotile was obtained by grinding chrysotile ore (Glove, Arizona) in a dry condition for 15 min in a high-energy vibratory ball mill. To evaluate the utility of different soil amendments often applied in the remediation of abandoned mines or polluted sites, we added 5 cm of a soil amendment to the surface: NPK fertilizer; a commercial organic, soilless potting medium; dehydrated cow manure; topsoil; or no amendment. We examined (1) how the presence of asbestos and the different soil amendments affected plant biomass, and (2) whether leachate collected weekly from the Cone-tainers^{TM}after watering contained asbestos fibers. We also collected the asbestos remaining in the containers at the end of the experiment; in the future, we will examine whether physical or chemical degradation of asbestos fibers occurred and varied among experimental treatments.

All soil amendments improved the aboveground biomass of both species compared to sand alone, but the fertilizer increased biomass the least. The effect of chrysotile on plant biomass depended on the soil amendment and species. Chrysotile increased biomass in the NPK fertilizer treatment for both plant species, perhaps because the chrysotile provided micronutrients. However, chrysotile decreased the biomass of *Brassica* in the presence of topsoil and decreased the biomass of *Sorghum*in the presence of all three amendments other than NPK. It is less likely that micronutrients are limiting in these other soil amendment treatments.

The movement of chrysotile fibers through the sandy substrate depended on the soil amendment but not the plant species. Measured by the 100.2 Method, a standard method used by EPA originally for chrysotile fibers >10 μm, chrysotile fibers >0.5 μm were found in large quantities when manure was used as a soil surface amendment but not with other soil amendments or no soil amendment. Fiber concentrations reached 100,000 MFL (millions of fibers per liter). As seen microscopically, the fibers collected in leachate were coated with organic matter particulates. The greatest concentrations were detected after 4 and 5 weeks of culture. The EPA threshold standard, applied to fibers >10 μm is 7 MFL. We found no measureable quantities of fibers >10 μm.

Data from asbestos contaminated sites corroborate our findings that asbestos can move in soil water. EPA has measured concentrations up to 0.51 MFL in test wells at a former industrial site in Pennsylvania, where chrysotile was used in manufacturing, and up to 100 MFL in surface water at an amphibole mine site in Montana. In both cases, the fiber size measured exceeded 10 μm. We found far higher concentrations in two test wells with 79 and 9 MFL at the Pennsylvania site when including fibers >0.5 μm.

Taken together, these studies provide evidence that asbestos fibers can contaminate ground water. Our experimental results suggest that adding manure as a soil amendment can increase the mobility of asbestos in soil. We, therefore, conclude that careful consideration must be given to how soil treatments affect the containment of the contaminant in question, even if the development of vegetation is facilitated.

Phosphorus Loading and Speciation Dynamics in Runoff Sediments Moving through a Eutrophic Watershed: Prospecting a Potentially-Recoverable Nutrient Resource

Presenting Author: Katie Gaviglio[1]

Andro Mondala[1], Shaun Shields[1], Steve Kaczmarek[2]

[1]Department of Chemical and Paper Engineering, Western Michigan University

[2]Department of Geosciences, Western Michigan University

17AIChE Annual Meeting, Minneapolis, MN

Western Michigan University

Phosphorus: A Critical Nonrenewable Resource For Food Security

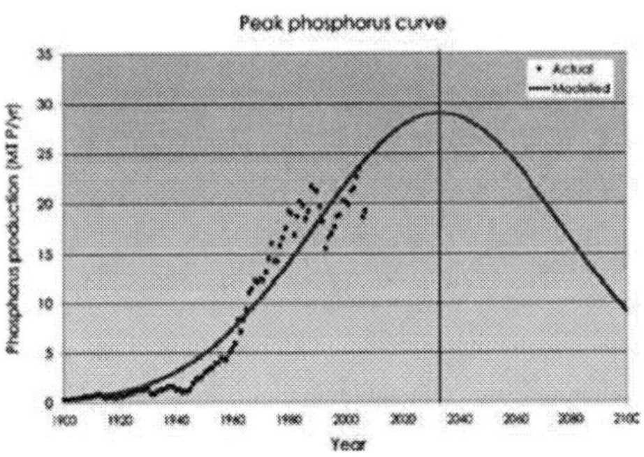

- As the world population increases, the global P demand is expected to grow
- Peak P predicted to occur in ~30 years (Cordell et al., 2009)

P is an nonrenewable resource

17AIChE Annual Meeting, Minneapolis, MN

Western Michigan University

Nonpoint Particulate P Losses in the Environment

Agricultural Runoff

Urban Runoff

Sediment and particulates transported to surface waters during precipitation and storm events

Represents 46% of the total P mined annually (Rittman et al., 2011)

Harmful Algal Bloom

Fish Kills

These processes lead to negative Impacts on communities and public health and also represents a loss of valuable resources

P is an over abundant Pollutant

17AIChE Annual Meeting, Minneapolis, MN

Western Michigan University

Effective P Management

P is an under abundant resources and an over abundant pollutant

1. Mitigates environmental impact
2. Regains the lost P resource for reuse
3. Reduces the demand on mined P ores

Artificial Wetlands

Ditches

1\. Effectiveness Varies 2. X 3. X

17AIChE Annual Meeting, Minneapolis, MN

Western Michigan University

Using P-Solubilizing Fungi (PSF)

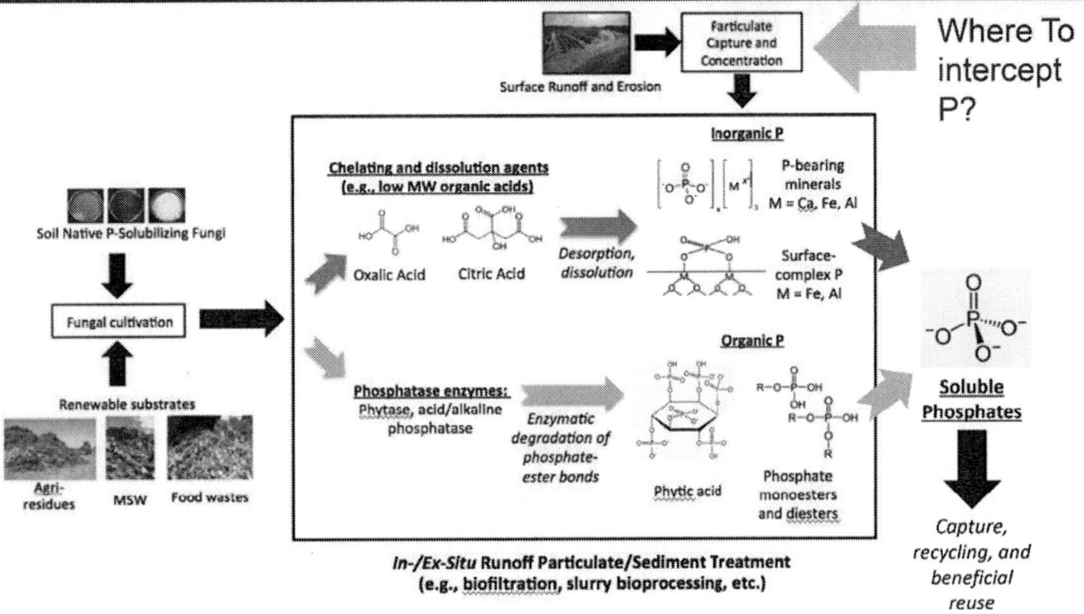

1. Mitigates environmental impact ✓
2. Regains the lost P resource for reuse ✓
3. Reduces the demand on mined P ores ✓

17AIChE Annual Meeting, Minneapolis, MN
Western Michigan University

Changes in P across watersheds

- Understanding how P changes is critical for…
  - Identification of optimum interception points for P recovery - "P Prospecting"
  - Ensuring robustness of PSF-mediated treatment
- Study Objectives
  - Characterize the P content and inorganic P speciation of runoff sediments from seven locations within the eutrophic Macatawa River Watershed
  - Observe how changes in P characterization correlate to site and sediment characteristics
  - Repeat study to observe temporal variation in P Characteristics
    - Examines optimal seasonal timeframe

17AIChE Annual Meeting, Minneapolis, MN
Western Michigan University

Approach

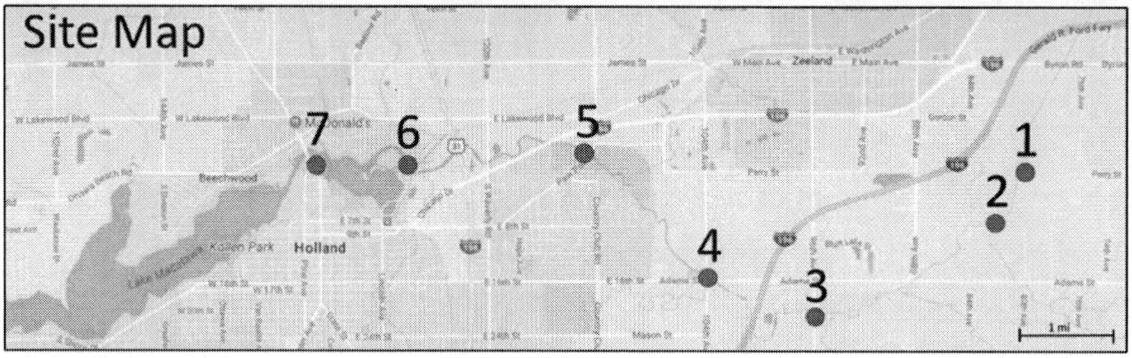

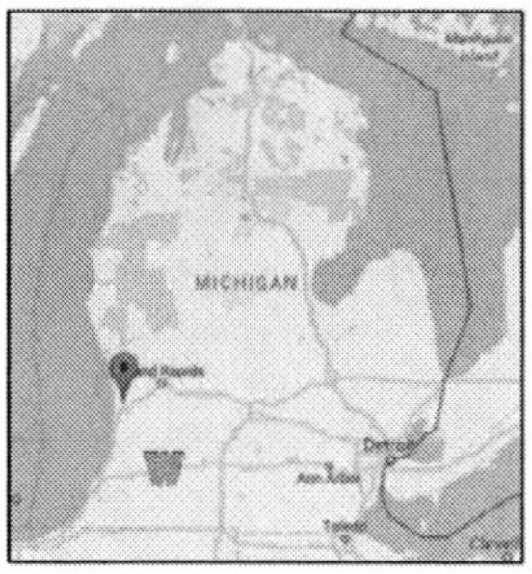

- During Collection
  - Qualitative observations of site and sediment characteristics
- Post Collection
  - Quantification of Total P, Inorganic P, and Organic P content
  - Quantification of Inorganic P speciation
    - Loosely-sorbed, Fe-bound, Al-bound, Ca-bound, and reductant soluble

17AIChE Annual Meeting, Minneapolis, MN

Western Michigan University

Qualitative Results: September 2016

Site	Surrounding Land Use/Type	Sediment description
1	Park; Expansive Wetlands	~100% silt with organic matter
2	Park; Deep Woodland	~90% sand and ~10% silt
3	Private Property; Agriculture and Residential	~90% sand and ~10% silt
4	Park; Woodland	~100% sand
5	Park; Woodland	~70% silt and ~30% clay
6	Park; Woodland, Maintained Landscape	~100% silt with organic matter
7	Commercial Area; Wetlands	~50% sand and 50% silt

17AIChE Annual Meeting, Minneapolis, MN

Western Michigan University

Results: Total, Inorganic, and Organic P Content

Sept. 2016

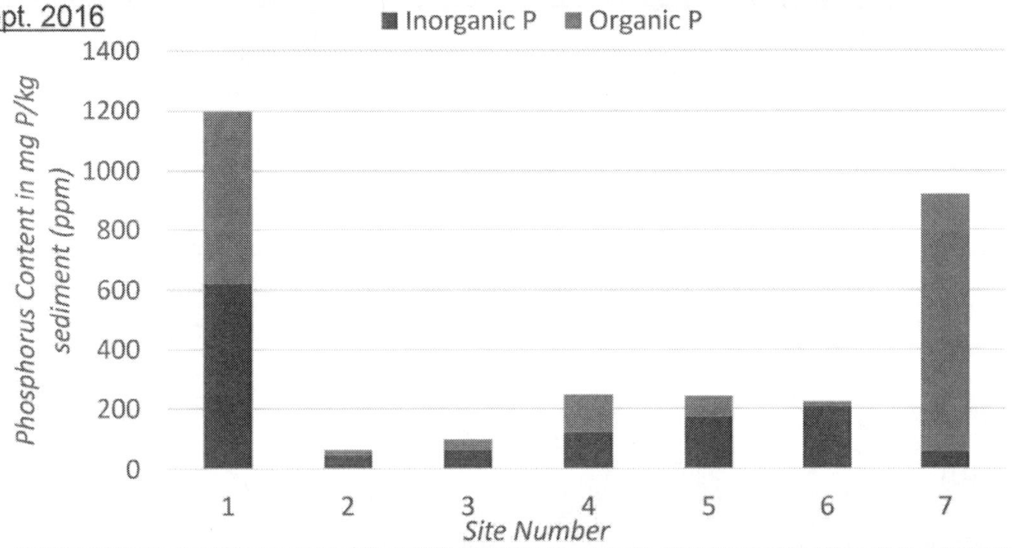

Sediment Type	~100 % Silt	~90 % Sand ~10% Silt	~90 % Sand ~10% Silt	~ 100% Sand	~70% Silt ~30% Clay	~100% Silt	~50% Sand ~50% Silt
Land Type	Expansive Wetlands	Deep Woodland	Agricultural/ Residential	Woodland	Woodland	Woodland/ Maintained landscape	Commercial Area/ Wetlands

Annual Meeting, Minneapolis, MN

WESTERN MICHIGAN UNIVERSITY

Results: Fractionation of Inorganic Species

Sept. 2016

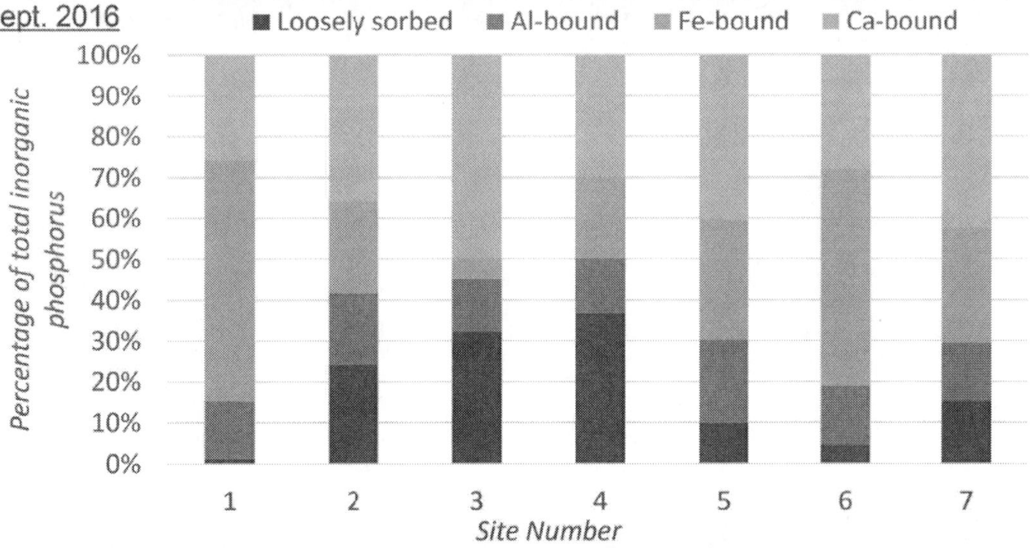

Sediment Type	~100 % Silt	~90 % Sand ~10% Silt	~90 % Sand ~10% Silt	~ 100% Sand	~70% Silt ~30% Clay	~100% Silt	~50% Sand ~50% Silt
Land Type	Expansive Wetlands	Deep Woodland	Agricultural/ Residential	Woodland	Woodland	Woodland/ Maintained landscape	Commercial Area/ Wetlands

Annual Meeting, Minneapolis, MN

WESTERN MICHIGAN UNIVERSITY

Qualitative Results: April 2017

Site	Surrounding Land Use/Type	Sediment description
1	Park; Expansive Wetlands	~100% silt with organic matter
2	Park; Deep Woodland	~90% sand and ~10% silt
3	Private Property; Agriculture and Residential	~90% sand and ~10% silt
4	Park; Woodland	~100% sand
5	Park; Woodland	~70% silt and ~30% clay
6	Park; Woodland, Maintained Landscape	~100% silt with organic matter
7	Commercial Area; Wetlands	~50% sand and 50% silt

17AIChE Annual Meeting, Minneapolis, MN

Western Michigan University

Results: Total, Inorganic, and Organic P Content

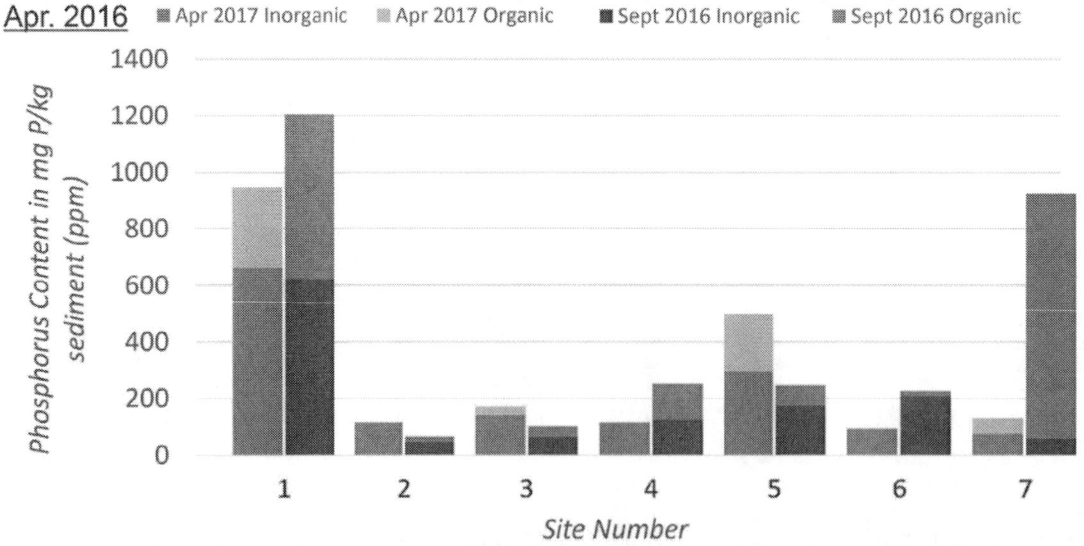

Sediment Type	~100 % Silt	~90 % Sand ~10% Silt	~90 % Sand ~10% Silt	~ 100% Sand	~70% Silt ~30% Clay	~100% Silt	~50% Sand ~50% Silt
Land Type	Expansive Wetlands	Deep Woodland	Agricultural/ Residential	Woodland	Woodland	Woodland/ Maintained landscape	Commercial Area/ Wetlands

17AIChE Annual Meeting, Minneapolis, MN

Western Michigan University

Conclusions

- For this particular watershed in Sept. 2016, the highest quantities of Total P correlated to the presence of wetlands at the collection site
- In Sept. 2016 there is no significant correlation between sediment type or land usage and the fractions of inorganic P species in the sediment at each site
- This study found that seasonal changes can redistribute P within the watershed and that the correlation of high P content to wetlands was not constant
- Of the sites tested, Site 1 is the most suitable for P recovery

17AIChE
Annual Meeting, Minneapolis, MN
WESTERN MICHIGAN UNIVERSITY

Acknowledgements

- WMU Office of the VP for Research. Faculty Research and Creative Activities Award (FRACAA) No. W2016-006

WESTERN MICHIGAN UNIVERSITY

- NSF MI-Louis Stokes Alliance for Minority Participation at WMU

NSF
MI-LSAMP
LOUIS STOKES ALLIANCE FOR MINORITY PARTICIPATION

- Outdoor Discovery Center Macatawa Greenway (ODCMG)

Outdoor Discovery Center Macatawa Greenway

17AIChE
Annual Meeting, Minneapolis, MN

Thanks for Listening! Questions?

Biosystems and Environmental Engineering Research
@ WMU Dept. of Chemical and Paper Engineering

17AIChE Annual Meeting, Minneapolis, MN
WESTERN MICHIGAN UNIVERSITY

17AIChE Annual Meeting, Minneapolis, MN
WESTERN MICHIGAN UNIVERSITY

Results: Total, Inorganic, and Organic P Content

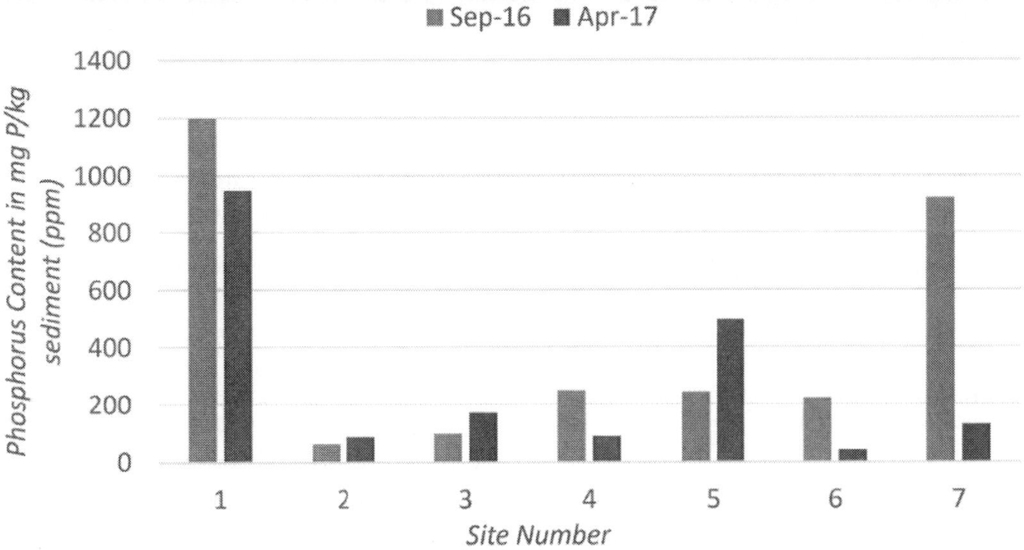

Sediment Type	~100 % Silt	~90 % Sand ~10% Silt	~90 % Sand ~10% Silt	~ 100% Sand	~70% Silt ~30% Clay	~100% Silt	~50% Sand ~50% Silt
Land Use *(Unchanged)*	Expansive Wetlands	Deep Woodland	Agricultural/ Residential	Woodland	Woodland	Woodland/ Maintained landscape	Commercial Area/ Wetlands

Annual Meeting, Minneapolis, MN

Western Michigan University

Picture of Sediments

17AIChE

Annual Meeting, Minneapolis, MN

Western Michigan University

494201 The Influence of Suspended Sediment on Electrochemical Remediation of Karst Groundwater

Abstract:

Electrochemical remediation is an innovative, sustainable, and promising approach for treatment of groundwater in complex terrains such as karst. Electroremediation offers high flexibility in generation and control of conditions needed for the contaminantsâ€™ transformation to harmless byproducts. However, performance of the treatment can be affected by the presence of suspended sediments, which appear in karst groundwater due to the direct recharge through sink holes and other areas. In this study, we evaluated the impacts of suspended sediments originating from Meff Cave (MC), Drop Lick Cave (DLC), and Crab Tree Spring (CTS) on electrochemical processes supporting the electro-Fenton reaction. We tested the production rates of *in situ* electro-generated hydrogen peroxide in the absence (control experiment) and presence of 2 g/L of each sediment sample (particles size <2 mm). The tests were conducted in the flow-through electrochemical reactor under current intensity of 90 mA and flow rate of 5 mL/min. Comparing to the control experiment, the hydrogen peroxide production decreased by 53% in presence of MC, 49% in presence of DLC and 71% in presence of CTS. The preliminary tests confirmed sorption of sediment particles at the electrodes, which is suspected to cause a decrease of the electrodesâ€™ active area and, therefore, limits the production of chemical species involved in hydrogen peroxide formation (oxygen and hydrogen from water electrolysis). Further, we are investigating other inhibition pathways, as well as the influence of additional redox-species from sediments such as Fe and Mn that may be responsible for hydrogen peroxide activation to hydroxyl radicals: a favorable pathway that supports creation of oxidizing conditions.

497132 Development of an Analytical Modeling Framework for Matrix Diffusion in Multi-Layered Systems

Abstract:

Matrix diffusion is the process whereby a contaminated low-permeability geologic layer(s) acts as a long-term source to a higher permeability aquifer. A quantitative understanding of matrix diffusion processes is necessary to predict the long-term fate and transport of a wide variety of groundwater contaminants in subsurface environments that include clays or silts. In this work, a modeling framework for the analytical solution of the one-dimensional diffusion equation was developed to quantitatively describe matrix diffusion in multi-layered systems. The modeling framework was then applied to evaluate the potential long-term effects of matrix-diffusion at a field site contaminated with chlorinated hydrocarbons. Finally, an inverse model was constructed and applied to estimate historic contaminant loading with respect to time given measured vertical concentration profiles within a low-permeability formation. This framework has potential applications to groundwater contaminant fate and transport, remedial design, and environmental forensics.

499064 Synthesis and Characterization of Fluorescent Polymers for the Detection of Polychlorinated Biphenyls

Abstract:

Polychlorinated biphenyls (PCBs) are a class of persistent organic pollutants (POPs) that have accumulated in the environment due to mass production in the early- to mid-1900s, and their highly stable structures inhibit natural remediation. Current experimental capture and remediation strategies rely heavily upon antiquated sensing methods (e.g. gas chromatography) for determination of an analyte's concentration in solution which are plagued by inaccuracy and irreproducibility. The extreme hydrophobicity of PCBs generates low sample concentrations and transfer losses that complicate traditional analyte sensing approaches. As a result, there is a significant need for a robust system that facilely detects PCB concentrations in solution to expedite capture and remediation technology advancements. One possible solution is the development of a polymeric system capable of coupling Förster resonance energy transfer (FRET) with selective, analyte-responsive domains. To this end, individual fluorophore conjugates are proposed here as tools for assessing the efficacy of affinity moieties and linker structures for the selective detection of individual PCB congeners. Ideal pairs will elicit sensitized emission in the presence of individual analytes and provide a colorimetric indicator of the analyte's concentration via fluorescence peak intensity shifts. High-performance liquid chromatography (HPLC), Fourier transform infrared spectroscopy (FTIR), and nuclear magnetic resonance spectroscopy (NMR) will be used to evaluate the success of linker synthesis and fluorophore-linker conjugation. Fluorescence spectrometry will also be employed to analyze pair fluorescence intensity and spectra shifts in response to the introduction of various PCB congeners and competing environmental contaminants. The performance of the standalone pairs will offer insights into the system's capability to selectively and accurately detect analytes in solution, and the pair synthesis approach will contribute a pathway to integrate the detection regime into a functional polymer system.

500249 The Benefits of Research Translation to Environmental Research: A Case Study Featuring Fate and Transport of per- and Polyfluoroalkyl Substances in New England

Abstract:

Adoption by the U.S. Environmental Protection Agency (EPA) of drinking water advisories of 0.07Î¼g/L for perfluorooctanoic acid (PFOA) and perfluorooctane sulfonic acid (PFOS) highlights growing concern about per- and polyfluoroalkyl substances (PFAAs) in the environment.Â PFASs have been used in industrial and consumer products and processes including fire fighting foams and fluoropolymer manufacturing, leading to environmental releases through use and disposal. Numerous communities in the New England region have recently identified PFASs in regional water resources including groundwater and surface water used as drinking water supplies. In response, the Brown University Superfund Research Program (SRP) Research Translation Core has worked in conjunction with SRP stakeholders to provide PFAS workshops, webinars, and analytical assistance to regulators and practitioners in the region. Through these events a series of knowledge gaps were outlined that represent priority challenges for management of PFASs in the environment. These knowledge gaps include tools for targeted assessment of potential regional sources of PFASs, improved understanding of PFAS physicochemical properties (e.g. vapor pressure) and implications for fate and transport, occurrence of PFASs in coastal ecosystems, and improved approaches for PFAS remediation. In response, the Brown SRP has developed a series of research objectives to address gaps in these key areas, and results from a subset of these efforts will be presented. Overall, research results have an immediate benefit to stakeholders actively managing environmental releases of PFASs and illustrate how environmental research can benefit from the bilateral communication between regulators/practitioners and scientific experts that is established during research translation efforts.

497367 Intensification of Cryogenic Lithiation-Borylation Though Use of a High Heat Transfer Loop Reactor

Abstract:

Lithium-hydrogen exchange is a common chemical transformation used to functionalize molecules, specifically to add boron to phenyl rings. One challenging class of these lithiations is the case where fluorine directs the lithium to the ortho position. These ortho-fluorine lithiated phenyl species can undergo highly exothermic decomposition through benzyne intermediates and are therefore run at deep cryogenic conditions. Not only does the intermediate need to be kept cold, but the subsequent borylation step also generates a substantial amount of heat that can promote the decomposition reaction. The challenge of handling the heat generation means operating this process requires good control of the temperature and rate of heat generation. In a batch process, this means slow addition of reagents at low temperatures. Changing operation from a batch to a continuous process presents the opportunity to increase the heat transfer rate and potentially support higher temperatures and faster mixing.

A small scale cryogenic flow chemistry platform was used to initially screen this reaction in a series of two tubular reactors.(1) At this milliliter scale, the lithiation was operated at -50 °C with the lithiated intermediate directly mixed with the boron source without degradation in a simple y-mixer. The reaction was further scaled up by approximately a factor of 3 to generate gram quantities of material and faced several challenges in the mixing section with respect to heat removal.(2) These were overcome by reconfiguring the mixing zones and better controlling heat removal from the reactor. These early results indicate that further scaling up of the process using the same configuration would present insurmountable problems.

We will discuss an additional 10x scale up capable of producing kilogram quantities of material over the course of a day. This required redesigning the borylation reactor to use a loop reactor with a heat exchanger inline to remove the heat of reaction rapidly. We will also show how careful design of the lithiation reactor would produce plug-flow performance even under laminar flow at the larger scale. The envisioned production scale continuous plant would have a 100x smaller footprint and reduction in potential chemical energy contained in the unstable intermediate. This would be safer and more economical than the equivalent batch process.

References:

1. Newby, J. et al. Design and Application of a Low-Temperature Continuous Flow Chemistry Platform. Org. Process Res. Dev. 2014, 18, 1211-1220.

2. Newby, J. et al. Reconfiguration of a Continuous Flow Platform for Extended Operation: Application to a Cryogenic Fluorine-Directed ortho-Lithiation Reaction. Org. Process Res. Dev. 2014, 18, 1221-1228.

497922 Design of Novel Microalgal Photbioreactor Using Computational Fluid Dynamics

Abstract:

The incessant rise of global CO_2 levels in the environment, depleting fossil fuel reserves and increased human & industrial activities are putting our ecosystem at risk. The search of renewable and sustainable options for generating fuels are on the rise. Microalgae is regarded as a green alternative owing to its several beneficial characteristics such as faster growth rates, CO_2 fixation abilities, etc. Moreover, microalgae has a tremendous potential for efficient CO_2 capture from point sources, biomass production for bio-product generation (e.g. biofuels, cosmetics, fodder and food etc.) and waste-water mitigation. However, one of the major bottlenecks in the microalgae-to-biofuel value chain lies in the growth of microalgae for commercial scale. Microalgae is cultivated in open as well as closed systems. However, the improved performance of a microalgal growth system is generally exhibited in closed photobioreactor which is governed by several physico-chemical factors such as light, pH, mixing, aeration, CO_2 etc. It is generally true that hydrodynamics plays a significant role by improving the mixing of nutrients, gases (improved mass transfer), avoiding the creation of dead zones, improved light penetrability and reduced sheer stress on the biological cells as well as reduced power consumption. In field experiments as well as computer simulation studies, the most common types of PBRs which are thoroughly investigated are bubble column and air-lift reactors. However, there is a scope to enhance flow patterns in PBRs using advanced CFD techniques. Computational Fluid Dynamics (CFD) enables determination of the hydrodynamics of the culture medium and mixing efficiency according to various PBR designs which is otherwise difficult to measure in field experiments. In this work Eulerâ€“Lagrangian model has been used to evaluate the turbulence mixing for two phase flow in airlift PBR using commercial CFD software. Three reactor configurations have been proposed with variation in draft tube length and type for assessing its effect on hydrodynamic characteristics. Parameters such as flow rate (Q), mean velocity (**u**), and mean Turbulent Kinetic Energy (TKE) were characterized for each configuration. The dimensions of the novel design were optimized using iterative design procedure. In all experiments, tap water was used as the liquid phase and was operated in a batch mode, while compressed air (CO_2 rich) was used as the gas phase and was operated in continuous mode. The hydrodynamics properties such as axial dispersion coefficient, mixing time and circulation time were calculated. The experimental results and CFD simulations were thoroughly compared for the novel PBR with that of conventional airlift reactor. In addition to performance evaluation of the novel PBR, a green microalga *Chlorella* sp. is cultivated in the PBR using air/flue gas and its growth characteristics were determined. The performance of the novel PBR was compared with that of the conventional airlift PBR. The results showed that the novel design had superior liquid circulation properties that resulted in a regular and alternative exposure to light and dark regions of the PBR. Also, CFD led to improved CO_2 mass transfer efficiency, biomass production, nutrient and light utilization profile.

501874 Process Intensified and Direct Production of Gasoline from Syngas

Abstract:

An Abstract Submission to 2017 AICHE Meeting

Session: Process Intensification By Enhanced Heat and Mass Transfer

Topical Area: Process Intensification

Process Intensified and Direct Production of Gasoline from Syngas

Xinquan Cheng[a], William R. Yantz Jr.[a], and Bruce J. Tatarchuk[a, b,*]

[a] Center *for Microfibrous Materials Manufacturing, Department of Chemical Engineering, 212 Ross Hall, Auburn University, AL 36849*

[b] *IntraMicron Inc., 368 Industry Dr., Auburn, AL 36832, USA*

* *Corresponding author. Tel.: +1 334 844 2023; fax: +1 334 844 2065; E-mail address: Tatarbj@auburn.edu*

Abstract

The increasing demand for liquid fuels draw people's attentions to the synthetic fuels. Currently, there are mainly two approaches to synthesize liquid fuels from syngas. One is Fischer-Tropsch synthesis (FTS) and another one is syngas-to-methanol first and followed up with methanol-to-gasoline (MTG). Those two approaches are well-known techniques for converting syngas to liquid fuels. However, FTS process produces a broad spectrum of products, which requires complex and expensive subsequent processes to upgrade the initial products in order to meet liquid fuel specifications. The MTG approach is a multiple step process. Typically, in the first reactor syngas is converted to methanol, and followed with methanol dehydration reactor to produce the intermediate product dimethoxyethane (DME), and then DME successively converted to hydrocarbons in the reactor packed with zeolite. It must be noticed that all those reactions are highly exothermic reactions, which producing extremely amount of heat during the reaction processes. Therefore, the reaction thermal management is always a big issue for syngas to liquid fuels.

Here, we present a single reactor system to directly convert syngas-to-gasoline (STG) by using a novel catalyst structure called Microfibrous Entrapped Catalyst (MFEC) structure. The MFEC structure was developed by our group, which is a catalyst network supported by micron-sized metal fibers with high thermal conductivity. By supporting small catalyst particles into a high voidage matrix of microfibers the catalyst is suitable for fixed bed applications without severe pressure drop or bed channeling encountered in small particle packed beds. Previous efforts on this unique catalyst structure have shown significant enhancement in intra-bed heat transfer and mass transfer.[1-4] Basically, we packed a large tubular reactor (34.0 mm I.D.) in three zones by using Cu MFEC. As we can see from Figure 1, the first zone (Zone-1) is a mixed catalyst zone for the production of DME from syngas. The second zone (Zone-2) is a relative inert zone mainly used to separate Zone-1 from Zone-3 since those two zones have different temperature. In order to convert all the unreacted methanol from Zone-1, we packed γ-Al_2O_3 in Zone-2 as a methanol dehydration catalyst. The third zone (Zone-3) is a zeolite zone packed with ZSM-5 to convert the DME to gasoline. This kind of multi-zone reactor design can separate the zeolite from the methanol synthesis catalyst, because the proximity of those two catalyst components is crucial factor for the product selectivity of STG process.[5] If those two components are close to each other, the lower olefins, which are the essential intermediates from methanol to aromatics, diffuse from the zeolite to the methanol synthesis catalyst, where they are easily hydrogenated to lower paraffins. It turns out that most of the products are short chain paraffins.[6]

However, with this simple multi-zone reactor design, the olefinic intermediates are preserved for subsequently oligomerization reactions. Under a typical reaction conditions (P=30 atm, GHSV=1000 h^{-1}, H_2:CO=1:1), the $C_5 - C_{10}$ hydrocarbon selectivity can reach 39.0 wt% with a relative high CO conversion (73.2 %). In addition, the Cu MFEC structure demonstrated a very low radial temperature gradient (around 5 °C) for a 34.0 mm I.D. reactor and insignificant amount of pressure drop. While with the same test conditions, the comparative packed bed reached a very high radial temperature gradient (around 140 °C) and significant amount of pressure drop. Considering these attributes and advantages, the Cu MFEC multi-zone reactor system is a very efficient process intensification technique for the direct production of gasoline from syngas.

Figure 1. The STG reactor design (all the catalysts particles entrapped in Cu MFEC in Zone-1, Zone-2, and Zone-3 are in 60 – 80 mesh)

References:

[1] M. Sheng, H.Y. Yang, D.R. Cahela, B.J. Tatarchuk, J. Catal. 281 (2011) 254-262.

[2] M. Sheng, H.Y. Yang, D.R. Cahela, W.R. Yantz Jr., C.F. Gonzalez, B.J. Tatarchuk, Appl. Catal. A-Gen. 445-446 (2012) 143-152.

[3] M. Sheng, D.R. Cahela, H.Y. Yang, C.F. Gonzalez, W.R. Yantz Jr., D.K. Harris, B.J. Tatarchuk, Int. J. Heat Mass Tran. 56 (2012) 10-19.

[4] X.Q. Cheng, H.Y. Yang, B.J. Tatarchuk, Catal. Today 273 (2016) 62-71.

[5] S. Sartipi, M. Makkee, F. Kapteijn, J. Gascon, Catal. Sci. Technol. 4 (2014) 893-907.

[6] K. Fujimoto, H. Saima, H. Tominaga, J. Catal. 94, (1985) 16-23.

487118 Modelling of Time-Dependent Interfacial Properties Due to Chemical Equilibrium Reactions in Demixed Fluid Systems

Abstract:

Modelling of time-dependent interfacial properties due to chemical equilibrium reactions in demixed fluid systems

Andreas Danzer, *Sabine Enders*

Karlsruhe Institute of Technology, Institute of Technical Thermodynamics and Refrigeration Engineering, Engler-Bunte-Ring 21, D-76131 Karlsruhe, Germany.

Email: andreas.danzer@kit.edu; sabine.enders@kit.edu

Key words: reactive extraction, impact of chemical reactions on interfacial properties, model calculations

Abstract:

Reactive extraction, the integration of reaction and extraction within one unit operation, is a promising process in a variety of applications. Compared with the conventional process of carrying out reaction and separation sequentially, the in-situ removal of products formed in the reaction zone can favorably influence conversion and selectivity, especially of equilibrium reactions. Together with a significantly reduced effort in the task of separation, this can result in major energy as well as capital cost savings. However, there are feasibility constraints resulting both from thermodynamics and chemistry. In addition, the process intensification leads to higher complexity and thus the need for research in terms of modelling productiveness. Because of that model calculations of hypothetically ternary liquid-liquid reactive extraction systems were performed in this work. The liquid-liquid phase equilibrium was calculated with the well-known, quite simple Margules activity model, the interfacial properties were modelled with the Margules equation in combination with the Density Gradient Theory [1-4] (DGT). It was assumed that the chemical reaction occurs in a A B manner, whereas the solvent C does not participate in the chemical equilibrium reaction. Furthermore, the assumption was made that mass transfer occurs at speed of infinity, so the chemical reaction is rate determining, in contrast to the reach of phase equilibrium. The empirical constants of the Margules equation $A_{i,j}$, the so called influence parameter κ of the DGT as well as the chemical equilibrium constant $K=k_1/k_2$ were specified due to the need to serve as a model. For real systems, $A_{i,j}$ must be fitted to experimental equilibrium data of the binary subsystems, the influence parameter κ has to be fitted at one experimental data point of the interfacial tension of each binary system with a miscibility gap and K is determined by the chemical reaction. It was found that the time, to reach chemical equilibrium in addition to the phase equilibrium, is strongly dependent on the shape of the miscibility gap as well as on the slope of the tie lines. Furthermore, the interfacial tension of the demixed system decreases over time, because of the fact that the product of the chemical equilibrium reaction act as a solubilizer. The slope of the interfacial tension over time strongly corresponds to the quotient of the reaction rate of forward and backward reaction. The new theoretical framework allows to the first time the calculation of time-dependent interfacial tensions and can further transferred to real systems.

Literature:

[1] J.W. Cahn, J.E. Hilliard, J. Chem. Phys. 28 (1958) 258.

[2] C.I. Poser, I.C. Sanchez, Macromolecules 14 (1981) 361.

[3] T. Grunert, H. Rudolph, S. Enders, Zeitschrift für Physikalische Chemie 227 (2013) 1.

[4] T. Grunert, S. Enders, Fluid Phase Equilibria 381 (2014) 46.

495073 Impact of Internals on Bubble Column Performance

Abstract:

Impact of internals in a bubble column reactor is a widely studied topic especially for the interesting hydrodynamics involved. Though numerous experiments have been performed utilizing horizontal, vertical internals and other packing materials, their direct, individual impacts on the bubble column performance is complex and not fully understood. Utilizing effective CFD simulations to recreate the flow conditions has become a necessity to understand these dynamics better. Based on the reaction, operating regime and flow conditions, it is possible to drive the back mixing, liquid recirculation and other parameters favorable to our requirements, by modifying the internal structures. We made an attempt to quantify the performance of various internal designs (taken from literature), for a given column, based on selected hydrodynamic parameters. With a better understanding of the flow patterns and their response to a number of internal structures, we take this forward and propose new designs which are tailored specifically for the reaction conditions and maintaining the flow in a particular regime (bubbly flow regime in our case) even at higher U_g thereby successfully delaying transition. We validate them by 3-d printing these designs and testing out their performance on a laboratory scale bubble column.

AUTHOR INDEX

AUTHOR INDEX

AUTHOR INDEX